ASCE STANDARD

ASCE/EWRI

69-19

Standard Guidelines for Managed Aquifer Recharge

ASCE
AMERICAN SOCIETY OF CIVIL ENGINEERS

EWRI
ASCE

PUBLISHED BY THE AMERICAN SOCIETY OF CIVIL ENGINEERS

Library of Congress Cataloging-in-Publication Data

Names: American Society of Civil Engineers, author.
Title: Standard guidelines for managed aquifer recharge, ASCE/EWRI 69-19 / American Society of Civil Engineers.
Description: Reston, Virginia : American Society of Civil Engineers, [2020] | Includes bibliographical references and index. | Summary: "Ongoing demand for water supplies and the need for water storage have led to Managed Aquifer Recharge (MAR) becoming an increasingly important component for both storage and supply in regional water planning and management. In recent decades, MAR projects have been successfully implemented around the world"– Provided by publisher.
Identifiers: LCCN 2020006154 | ISBN 9780784415283 (paperback) | ISBN 9780784482179 (adobe pdf)
Subjects: LCSH: Artificial groundwater recharge–Standards–United States.
Classification: LCC TD404 .A48 2020 | DDC 627/.56021873–dc23
LC record available at https://lccn.loc.gov/2020006154

Published by American Society of Civil Engineers
1801 Alexander Bell Drive
Reston, Virginia 20191
www.asce.org/pubs

This standard was developed by a consensus standards development process that has been accredited by the American National Standards Institute (ANSI). Accreditation by ANSI, a voluntary accreditation body representing public and private sector standards development organizations in the United States and abroad, signifies that the standards development process used by ASCE has met the ANSI requirements for openness, balance, consensus, and due process.

Errata: **Errata, if any, can be found at https://doi.org/10.1061/9780784415283.**

ISBN 978-0-7844-1528-3 (paper)
ISBN 978-0-7844-8217-9 (PDF)

Manufactured in the United States of America.

25 24 23 22 21 20 1 2 3 4 5

ASCE STANDARDS

In 2016, the Board of Direction approved revisions to the ASCE Rules for Standards Committees to govern the writing and maintenance of standards developed by ASCE. All such standards are developed by a consensus standards process managed by the ASCE Codes and Standards Committee (CSC). The consensus process includes balloting by a balanced standards committee and reviewing during a public comment period. All standards are updated or reaffirmed by the same process at intervals between five to ten years. Requests for formal interpretations shall be processed in accordance with Section 7 of ASCE Rules for Standards Committees, which are available at http://www.asce.org/uploadedFiles/Technical_Areas/Codes_and_Standards/Content_Pieces/asce-rules-standards-committees.pdf. Errata, addenda, supplements, and interpretations, if any, for this standard can also be found at http://ascelibrary.org/.

The provisions of this document are written in permissive language and, as such, offer to the user a series of options or instructions, but do not prescribe a specific course of action. Significant judgment is left to the user of this document.

This standard has been prepared in accordance with recognized engineering principles and should not be used without the user's competent knowledge for a given application. The publication of this standard by ASCE is not intended to warrant that the information contained therein is suitable for any general or specific use, and ASCE takes no position respecting the validity of patent rights. The user is advised that the determination of patent rights or risk of infringement is entirely his or her own responsibility.

A complete list of currently available standards is available in the ASCE Library (https://ascelibrary.org/page/books/s-standards).

CONTENTS

PREFACE

Ongoing demand for water supplies and the need for water storage have led to managed aquifer recharge (MAR) becoming an increasingly important component for both storage and supply in regional water planning and management. In recent decades, MAR projects have been successfully implemented around the world. However, many water resources professionals remain unaware of the benefits, techniques, standard practices, and applications of MAR. Over the past several decades much has been learned regarding how to design and implement a successful, cost-efficient MAR project and what pitfalls exist that prevent MAR success. Although there are extensive resources about various specialized aspects of MAR in the scientific literature, the US water resources sector has lacked an up-to-date, comprehensive document describing the state of the practice for MAR projects.

In 2001, the American Society of Civil Engineers' Environmental and Water Resources Institute (ASCE\EWRI) published the first standard guidelines on aquifer recharge as EWRI/ASCE 34-01, *Standard Guidelines for Artificial Recharge of Ground Water.* As more MAR projects were developed and constructed, it became evident that the original standard needed to be modernized and upgraded to provide additional details on the planning, design, construction, and operation of MAR projects. In 2005, the ASCE/EWRI formed the Guideline Development Subcommittee for MAR. The purpose of this subcommittee was to provide a thorough and up-to-date document that describes the state of the practice for MAR projects. Beginning with the withdrawn standard (EWRI/ASCE 34-01), the subcommittee restructured the document to include details on planning, design, construction, operation, and monitoring of MAR projects, along with background information on groundwater and MAR concepts. This standard has been designed to meet the needs of water resources planners and stakeholders during the initial evaluation and planning phases, and the needs of engineers, hydrologists, and other professionals for standardization of MAR practices from conceptualization to operation.

ACKNOWLEDGMENTS

The ASCE and its Environmental and Water Resources Institute (EWRI) gratefully acknowledge the devoted efforts of the Managed Aquifer Recharge Guideline Development Subcommittee of the Managed Aquifer Recharge (MAR) Standards Committee of the Standards Development Council of EWRI. This standard was prepared by the MAR Standards Committee through ASCE's consensus standardization process in compliance with the ASCE Rules for Standards Committees and the procedures of ASCE's Codes and Standards Committee. ASCE's standardization process is accredited by the American National Standards Institute (ANSI).

The Managed Aquifer Recharge Guideline Development Subcommittee comprises individuals from many backgrounds, including consulting engineering, education, and government. Those individuals who served on the Managed Aquifer Recharge Guideline Development Subcommittee and actively participated in the drafting of ASCE Standard 69-19 are

Gordon McCurry, *Corresponding Editor and Co-Chair*
Peter Barkmann, *Co-Chair*
Doug Bartlett
Adam Hutchinson
Dennis McGrane
Stephanie Moore
Zhuping Sheng
Phyllis Stanin

Other current or former members of the Managed Aquifer Recharge Guideline Development Subcommittee who participated in review and balloting of the draft document include

Courtney Brand
Peter Dillon
Devin Galloway
Conrad Keyes, Jr.
Keith Prince
R. David G. Pyne
Ben Willardson
Greg Woodside

CHAPTER 1
INTRODUCTION

The growing demand for water supplies has led to an increasing need to store water to help meet periods of high demand. Aquifers form natural reservoirs that in many locations are capable of storing tremendous volumes of water. Worldwide, aquifers are being used more frequently for water storage for many reasons, including their proximity to sources of supply or demand, fewer environmental impacts relative to surface reservoirs, low evaporative losses and relatively low capital costs compared with surface storage in reservoirs, and their ability to be developed incrementally as needs arise. Storing water in aquifers can be as simple as enhancing natural recharge processes, or it can entail constructing a complex of recharge facilities.

Actively storing water in aquifers often involves the combined or conjunctive use of surface water and groundwater supplies, wherein water is recharged underground during times of surplus surface water for later withdrawal during times of surface water shortage. Implementation requires comprehensive planning, design, construction, ongoing monitoring, and active management to make best use of the combined surface water and groundwater resources. Because the recharge and storage of water in aquifers can be accomplished by a variety of intentional processes, it is referred to in this standard guidelines and elsewhere as managed aquifer recharge, or MAR. Some use the term ASR (aquifer storage and recovery) to denote the concepts referred to herein as MAR; these terms can be used interchangeably, although the authors of this standard guidelines prefer the term MAR.

1.1 PURPOSE

The purpose of this document is to provide a set of standard guidelines describing the activities required to implement a MAR project. This standard supersedes but builds upon *Standard Guidelines for Artificial Recharge of Ground Water* (ASCE 2001).

MAR projects are inherently interdisciplinary in their design, construction, implementation, and maintenance. This standard has been developed so that project personnel interested in, but perhaps with limited expertise with MAR, should benefit from understanding how to plan, design, construct, operate and maintain, and close MAR facilities based on the work of others.

1.2 SCOPE

MAR projects progress through a series of phases that normally begin with initial planning activities and then continue with design, construction, operation, and finally closure of recharge facilities. This standard follows the progression through these phases, describing the key activities involved in each phase, and includes the economic, environmental, and legal (water rights, laws, and regulations) considerations, and site evaluation and field-testing procedures. Figure 1-1 illustrates the five primary phases of MAR project progression described in this standard. Greater detail of the activities is provided in subsequent sections. As shown by the arrows on Figure 1-1, at many stages of a MAR project, the process may step back to earlier phases as new data are obtained or conditions change.

MAR may be accomplished either by applying water to the ground surface, or shallow subsurface, for infiltration into the underlying aquifers or by injecting it directly into aquifers through wells. MAR projects can be regional in scale, targeting entire aquifer systems or local projects addressing the needs of individual wells or water supply systems. This standard has been developed to cover situations that may occur with many different types of MAR projects in many different types of hydrogeologic settings. This standard also applies to any project regardless of size.

1.3 ORGANIZATION OF THIS STANDARD GUIDELINES

Following the presentation of background material, this standard follows the progression of MAR activities outlined in Figure 1-1. The document begins with an introduction and overview in this chapter. Chapters 2 and 3 provide basic descriptions of groundwater and the key MAR concepts; these chapters are intended for readers with limited training and experience in MAR. MAR implementation begins with planning and evaluation, described in Chapter 4. Chapter 5 presents facility design. Chapters 6 through 8 present construction, operations, and closure, respectively.

Data collection is an essential component of each phase of a MAR project (Figure 1-1), and the level of effort varies during each phase, depending on specific project goals. Chapter 9 presents data collection methods and special considerations.

A glossary of key terms is found in Appendix A. Appendix B defines notations and symbols used in the standard. Appendix C lists MAR regulations in the United States. Case studies that provide examples of MAR projects are summarized in Appendix D. The dimensions and quantities used in this document are in International System of Units (SI units) followed by conversion to US Customary units in parentheses per the ASCE Rules for Standards Committees (2016).

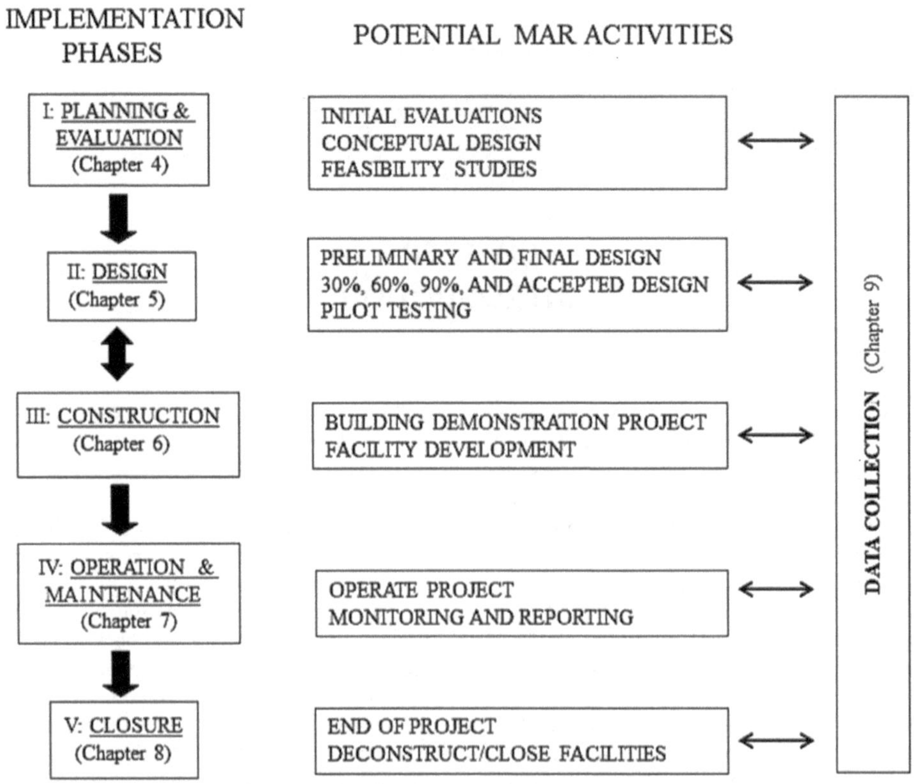

Figure 1-1. Implementation phases and potential activities of a MAR project.

CHAPTER 2
GROUNDWATER FUNDAMENTALS AND OCCURRENCE

This chapter includes a brief description of the hydrologic cycle that pertains to groundwater occurrence, recharge, and quality. Additional information on these topics can be found in a number of references, including Bouwer (1978), Driscoll (1986), Fetter (2001), Heath (1989), Pettyjohn et al. (1979), Todd and Mays (2005), and Winter et al. (1998).

2.1 HYDROLOGIC CYCLE

Groundwater is an important component of the world's total water supply. This resource, upon which all life depends, also includes water in the atmosphere, water in the oceans, surface water in streams and lakes, and ice in glaciers and icecaps. Water continually moves from one component to another in the system, which is referred to as the hydrologic cycle or the water cycle, as illustrated in Figure 2-1. Most of the world's water, approximately 97%, is salt water and not available for human use without treatment. Of the remaining freshwater supplies, approximately 69.6% is in glaciers and icecaps; 30.1% is groundwater; and the remaining 0.3% is found in surface water in rivers, streams, and lakes and in the atmosphere (Shiklomanov 1993).

2.2 GROUNDWATER OCCURRENCE AND USE

Groundwater is that portion of subsurface water filling interstitial pore spaces, voids, or fractures in soils, sediments, or rocks below the land surface. It is at a pressure equal to or greater than atmospheric pressure so that the water can flow into a well or other open hole. Geologic formations that contain groundwater and are sufficiently permeable to yield usable quantities of water from wells are called aquifers. Aquifers are typically composed of sand, gravel, sandstone, limestone, or fractured bedrock and are classified as either confined or unconfined. The top of unconfined aquifers is denoted by the water table surface, which can move up or down as water is added or withdrawn from them. Confined aquifers are sandwiched between relatively impermeable layers called aquitards. When tapped by wells, water in a confined aquifer rises above the physical top of the aquifer and defines the potentiometric surface for that aquifer. A third and common type of aquifer is labeled semiconfined, which is an aquifer in which the deposits above and below the aquifer are somewhat permeable, allowing leakage from the aquifer into adjacent layers. In this document, semiconfined and confined aquifers are lumped together for discussion purposes and referred to as confined aquifers.

Unconfined aquifers yield water to pumped wells by a lowering of the groundwater table (drawdown) around wells and associated gravity drainage and dewatering of pore spaces. Confined aquifers can yield water to pumped wells by expansion of water caused by the release of pressure; by compaction of clay layers or lenses in the aquifer caused by lowering of the pore water pressures; and, where such aquifers exist, by drawing water from storage in the unconfined lateral extensions of the aquifer. Because of these different mechanisms, water yielded by unconfined aquifers per unit drop of groundwater table is orders of magnitude higher than that yielded by confined aquifers per unit drop of potentiometric surface. By the same token, much more water can be stored in unconfined aquifers per unit of rise of the water table than in confined aquifers per unit rise in the potentiometric surface. This is an important consideration when evaluating managed aquifer recharge (MAR) projects.

Groundwater is an important source of water for many types of uses. As of 2015, more than 98% of the domestic water use and approximately 39% of the public water supply systems use groundwater as the source. Groundwater is also the source of more than 48% of the water used for agriculture and is an important source for livestock, mining, and industrial uses (Deiter et al. 2018).

Even though groundwater use is widespread, typical consumers have only a vague concept of where their water originates; how it is produced, treated and delivered; and what steps are taken to ensure its safety and sustainability for use. Until relatively recently, the public's perception of an aquifer often was one of an underground stream or lake that produces a generally unlimited, safe, and dependable source of water supply. That perception has changed as reports of groundwater contamination from leaking fuel storage tanks, municipal landfills, hazardous waste handling and disposal facilities, and pollution from countless other sources have been publicized. Similarly, with increased reports of declining water tables and the resulting shortages in supply, the public now has a greater awareness that there are limits to the supply of groundwater and the ability of natural recharge to sustain the demands being placed on aquifers throughout the country and the world.

2.3 GROUNDWATER MOVEMENT

Groundwater is continually in a state of movement, although at rates much lower than for surface water, from areas of recharge to areas of discharge. Groundwater residence times between its points of recharge and discharge can be on the order of days for wet weather springs; months to years for shallow groundwater in high rainfall areas; decades to centuries for deeper groundwater in humid areas; and thousands to tens of thousands of years for deep bedrock groundwater in dry climates. Groundwater from the Great Artesian Basin in Australia, for example, has been dated at several million years old. The prolonged residence time for most groundwater means that, on human timescales, most of it can be considered to be in storage. The concepts of groundwater recharge, storage, and discharge are described more fully in the following sections.

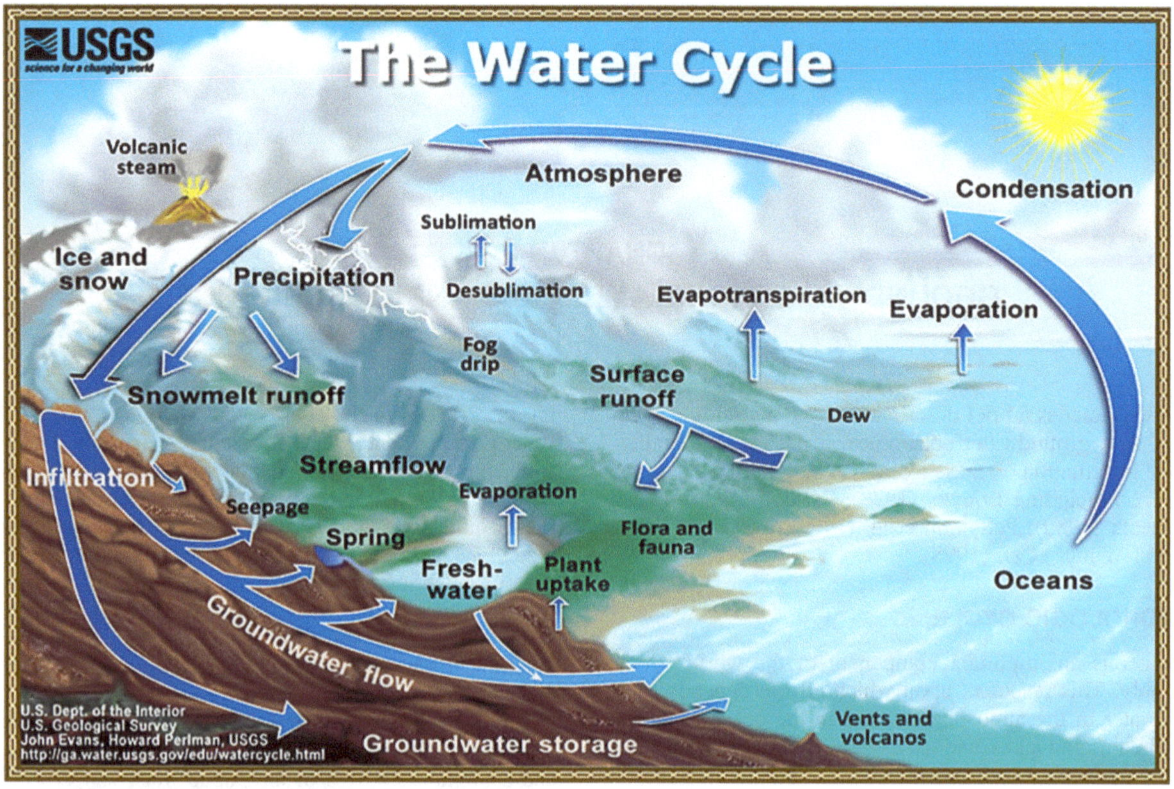

Figure 2-1. The hydrologic cycle.
Source: http://water.usgs.gov/edu/watercyclehi.html.

2.3.1 Groundwater Recharge. Unconfined aquifers are recharged primarily by deep percolation from the land surface or near the land surface. Long-term natural recharge rates for unconfined aquifers at the basin and regional scales are on the order of 20% to 30% of the average precipitation for humid climates, based on a variety of calculation methods. Many of these basins have relatively flat topography and permeable soils that produce little or no surface runoff. In contrast, for arid to semiarid climates [precipitation less than 20 cm/year (8 in./year)], recharge can be less than 1% of precipitation (Delin and Risser 2007). Most of the natural recharge in dry climates may actually come from leakage through the beds of flowing streams and from recharge along mountain fronts. Sources of artificial recharge to unconfined aquifers include infiltration of excess water applied to irrigated fields and landscaping, leakage from canals and artificially constructed reservoirs, and MAR projects. Unconfined aquifers can also be recharged by flow from deeper confined aquifers.

Confined aquifers are recharged naturally at their outcrops, where they are unconfined and can be recharged by leakage through their confining layers or through geologic faults from adjacent aquifers. Sources of artificial recharge to confined aquifers include leaky wells, mine shafts, and other borings that penetrate a confining layer and adjacent aquifer. The rate of natural recharge for confined aquifers is very small and on human timescales is often negligible; thus, the water is considered essentially nonrenewable in most confined aquifers.

2.3.2 Groundwater Storage. Aquifer storage potential and retention characteristics are determined largely by the nature and extent of the aquifer material storage coefficients, existence of fractures or faults, and outcrop formations near a potential MAR site. Generally, aquifers that underlie potential sites are extensive and may have experienced decreased storage (lower groundwater levels) because of groundwater extraction, such that recharge is not initially limited by available storage space. In other areas using well recharge, storage may be obtained by displacement of native water within a confined aquifer and some unconfined aquifers. Such native water may be of varying quality (ranging from fresh, brackish, or high in nitrates).

The use of a groundwater basin can be related to that of a surface reservoir where portions of the storage capacity are below the spillway elevation and thus are unrecoverable, and other portions might be reserved for flood control, water supply, and other uses. The usable portion of an aquifer may be further constrained by upper and lower water level regulations, by the zones penetrated by extraction wells, by potential for subsidence or seawater intrusion, or by water quality problems. The amount of usable storage has to be carefully defined and should not infringe on volumes needed for other uses.

For an unconfined (water table) aquifer, the potential amount of storage in a groundwater basin that can be used for MAR is the amount between the upper and lower operating water level limits of an aquifer. The upper operating water level limit is the highest level that water should be allowed to rise in an aquifer. This level is normally feet to tens of feet below the ground surface to reduce the risk of flooding of nearby structures, affecting the stability of foundations, water losses to vegetation, and discharge to surface drainages. The lower operating water level limit is commonly the base of the aquifer but may be other depths, such as the base of the water of a certain quality (such as the base of the freshwater zone) or the deepest a well can economically be pumped.

The amount of storage available in a confined aquifer is generally limited by allowable pressures to avoid fracturing the aquifer or aquitards and creating flowing artesian conditions in nearby wells. In cases where a confined aquifer has an outcrop area, the amount of storage available is limited to that available at the outcrop (recharge) area. For a confined aquifer containing nonpotable water, storage is obtained by displacement of native water around the well.

Field investigations are needed to quantify aquifer storage potential. The objective of the evaluation should be to determine the current amount and possible storage capacity of the groundwater basin or to confirm that the storage capacity exceeds project requirements. An additional objective is to determine the values of the parameters controlling the movement of groundwater within the basin and at possible recharge sites. Aquifer extent and other geologic features, such as lenses within the aquifer and fractures, faults, and other characteristics within adjacent strata, can be determined through analysis of drillers' logs, geophysical exploration, physical and chemical laboratory tests, well pumping tests, and observation of subsurface flow using tracers. Porosity, specific yield, and storage coefficient for prediction of storage potential can be estimated through the use of laboratory tests, such as grain size analyses and permeameter tests, and from in situ tests, such as aquifer pumping and infiltration tests, as described in ASTM Standards (D425, D653, D4043, D4044, D4050, D4104, D4105, and D4106). Chapter 9 discusses data collection activities in detail.

Formations containing natural openings or cavities may have high rates of recharge, causing water to flow quickly to other areas. This recharge may occur in limestone or in fractured rock deposits, where flow through secondary porosity in the form of horizontal and vertical channels, fractures, and voids can be the primary transport mechanism for groundwater. These formations are typified by small storage coefficients and discrete zones of high hydraulic conductivity. Fault fractures in both consolidated and unconsolidated materials may provide either a pathway or a barrier to movement of groundwater. Fault barriers may cause the separation of a basin into completely or partially isolated subbasins, allow vertical movement of water between aquifers, slow down water movement in an aquifer, or completely stop migration of water. Field investigations for aquifers located in faulted geologic settings are critical in identifying the feasibility of a MAR project.

Recharge to basalt bedrock has been successful in the Pacific Northwest. Case Study 12 in Appendix D describes wells that were completed in Miocene Age continental flood–basalt flows of the Columbia River Basalt Group. These wells use the aquifer storage provided by thick interflow zones between basalt flows. The city of Beaverton, Oregon, has installed three aquifer storage and recovery (ASR) wells that provide additional capacity for meeting peak demand for the city. Total operational pumping of 30,500 m^3/day [8 million gal./day (MGD)] is achieved from the ASR wells during the dry summer months. This pumping has provided the city of Beaverton with significant cost savings.

2.3.3 Groundwater Discharge. Groundwater can discharge (flow out of an aquifer) through both natural and artificially constructed mechanisms. Groundwater can discharge under natural conditions to surface water bodies, such as streams, lakes, and springs, to plants, to the atmosphere, and to other aquifers. Groundwater discharge to streams provides their baseflow and is responsible for keeping many streams flowing, even during prolonged periods when there is no precipitation. Groundwater discharge to plants and the atmosphere occurs when the water table surface is close to or at the land surface, where plant roots can access the groundwater, and where groundwater evaporates. Artificially constructed mechanisms for groundwater discharge include drainage ditches and pumping by wells.

2.4 GROUNDWATER QUALITY

Groundwater is in direct contact with the soil or rock through which it travels and dissolves some of the solid materials along the way. The longer its contact time between areas of recharge and discharge, the more minerals groundwater can dissolve. Generally, dissolution of soluble minerals and ion exchange cause groundwater to evolve from a dilute calcium bicarbonate type water to a more concentrated sodium chloride or calcium chloride type water.

Rainwater, the natural source of much groundwater, usually is of good quality with total dissolved solids (TDS) concentration ranging from about 1 to 50 mg/L (parts per million; ppm) with an average of about 10 mg/L. The TDS concentration of groundwater is often many times higher than this because of the dissolution processes, and the concentration increases with increasing time underground. Groundwater unaffected by human activity maintains a relatively consistent temperature and composition throughout the year, although these vary greatly from location to location, depending on the elevation, latitude, depth to the groundwater table, and composition of the aquifer materials.

Unpolluted or pristine groundwater can contain a variety of naturally occurring substances that may be harmful if present in high concentrations, such as arsenic, selenium, uranium, and heavy metals. Some confined and semiconfined aquifers contain at least one unwanted water quality constituent for stored water that is eventually recovered from the aquifer. Groundwater also typically contains 0.2 to 0.7 mg/L total organic carbon (Thurman 1979), which can form undesirable disinfection by-products when chlorinated. It also can contain viable microorganisms, even at great depths. Thus, pristine groundwater, although in many cases suitable for drinking without any treatment, is by no means pure and should be tested before being consumed.

CHAPTER 3
MANAGED AQUIFER RECHARGE CONCEPTS

This chapter provides a detailed discussion of managed aquifer recharge (MAR) concepts, including purposes and objectives, components of MAR systems, sources of recharge water, pretreatment, recharge methods, subsurface storage, recovery, post-treatment, and end uses that should be considered when undertaking MAR projects. As noted in Chapter 1, some people use the term ASR (aquifer storage and recovery) to denote the concepts referred to herein as MAR.

3.1 PURPOSES AND OBJECTIVES

The primary purposes of MAR projects are to (1) store water for later use, and (2) combat adverse groundwater conditions, such as overdevelopment of groundwater, subsidence, or saltwater intrusion. In general, objectives for MAR systems fall into one or a combination of the following categories:

- Water supply management,
- Water quality management,
- Aquifer restoration,
- Environmental mitigation and/or protection, and
- Meeting legal obligations.

The next section describes the components of a MAR system, which are the physical elements of the complete system. For the purposes of this standard, the components of the system are separate and distinct from the phases of a MAR project, which describe the lifecycle of a MAR project, from start to finish, shown in Figure 1-1. Chapter 4 on planning describes specific objectives for MAR in detail.

3.2 COMPONENTS OF A MAR SYSTEM

There are seven basic components of MAR projects: (1) sources of recharge water, (2) pretreatment, (3) recharge, (4) subsurface storage, (5) recovery, (6) post-treatment, and (7) end uses of recovered water.

Figure 3-1 depicts each component graphically for subsurface and surface recharge facilities. The term capture zone shown on the figure is meant to reflect sources of recharge water or source water discussed herein. The source water, recharge, and storage are essential components of any MAR project. Depending on the purpose of a particular MAR project, the pretreatment, recovery, post-treatment, and end uses also may be considered essential. For example, for a MAR project whose primary purpose is to combat saltwater intrusion, some of the stored water may never be recovered, yet the project is still considered a MAR project because it delivers one type of source water to a target aquifer for storage. Each of these components is described briefly in the remainder of this chapter.

3.3 SOURCES OF RECHARGE WATER

Virtually any available water source may serve as a source of water for aquifer recharge. The basic categories of source water for a MAR project include surface water, groundwater, reclaimed water, and desalinated brackish water or seawater. Potential surface water sources include local and imported water, stormwater runoff and harvested rainwater, and irrigation return flow. In some cases, groundwater from one aquifer is the source of water that might be stored in another aquifer, or in the same aquifer at a different location. Reclaimed water sources include wastewater effluent (and, in some cases, stormwater) that has been treated to meet relevant water quality standards for use in a MAR project. Treated wastewater, stormwater, and desalinated water are increasingly being used as sources for MAR systems (Bouwer 1997, CSIRO 2016, NRC 2008, NRMMC et al. 2009, USEPA 2016).

Many projects use a combination of water sources. For example, water sources used for recharge at the Montebello Forebay Spreading Grounds (Los Angeles Coastal Plain, California) have included stormwater, in-stream river flows, imported water, tertiary-treated wastewater, and advance-treated recycled water. Over time, the use of recycled water has become more important to decrease the reliance on imported water. Currently, tertiary-treated wastewater and recycled water are blended to meet water quality objectives. See Case Study 6 (Appendix D) for more details.

3.4 PRETREATMENT

The water quality of the source water must be evaluated with respect to the project purpose, the quality of water in the target aquifer, and the recharge method. Depending on the type of recharge process and the physical, chemical, and biological character of the recharge water, pretreatment may be necessary to make the source water suitable for recharge to the target aquifer.

Pretreatment processes range from the use of the soil profile and aquifer as a treatment medium (known as soil aquifer treatment or SAT), to complex physical and chemical processes used in water treatment plants before recharge. Generally, water recharged with wells requires water of higher quality than that recharged by surface infiltration. In all cases, the particulate matter of source water must be tested to evaluate the need to remove particulate matter before recharge. This testing is important to reduce clogging in recharge basins and particularly in the aquifer around recharge wells.

The water used for subsurface recharge needs to be treated to remove suspended materials and entrained air to remove nutrients and biodegradable organic carbon and to disinfect or otherwise inactivate microorganisms. These constituents should be removed to reduce the potential for physical and biological

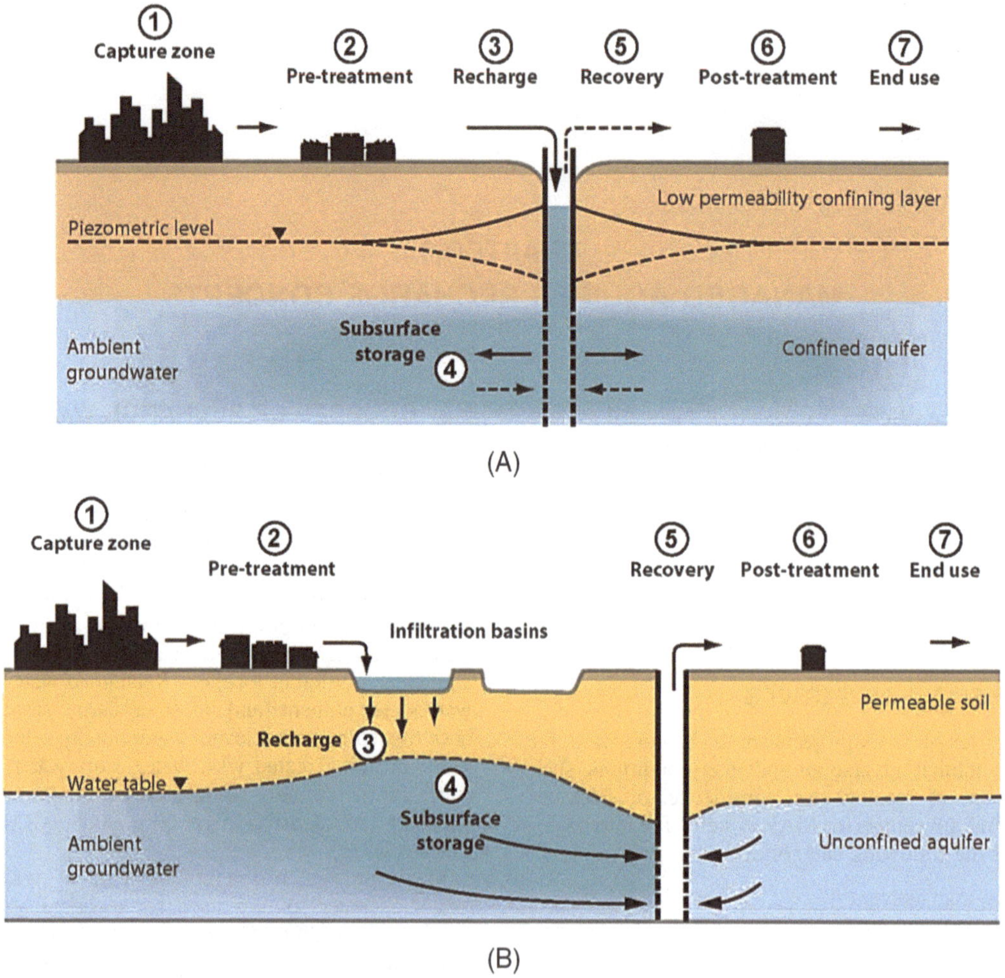

Figure 3-1. Components of a MAR system: (A) aquifer storage and recovery system storing water in a confined aquifer; (B) soil aquifer treatment system infiltrating water to an unconfined aquifer.
Source: NRMMC et al. (2009), Figure 2.1.

clogging of the aquifer or of the filter material around the screen or open portion of the well, and to prevent entry of pathogens, nonbiodegradable organic chemicals, or toxic chemicals into the aquifer.

The chemistry of the water to be recharged should also be compatible with the aquifer materials and groundwater chemistry to avoid precipitation and formation of colloids, or reaction with aquifer materials that may reduce aquifer permeability. Swelling of clays for freshwater storage in brackish aquifers can clog wells. Differences in pH and redox potential (oxidation reduction potential or ORP) between the recharge water and the aquifer can also promote chemical reactions that may affect permeability of the aquifer and filter pack surrounding a well. Even then, recharge wells need to be periodically pumped and possibly redeveloped or rehabilitated to alleviate the effects of clogging and to maintain satisfactory recharge rates. Pumping or backflushing to control well clogging may be needed every few days to every few months.

In some instances, regulatory agencies have developed additional pretreatment requirements for reclaimed water and other waters of poor quality before such waters can be used for recharge. Regulatory requirements for aquifer recharge must be understood and then implemented when designing pretreatment systems. For example, the required residence time of recharged water between recharge and recovery may be several months, providing an opportunity for additional soil aquifer treatment (Sheng 2005, Case Study 11 in Appendix D). Regulatory requirements may exist at the federal, state, and local levels so some research on this in necessary, as discussed in Section 4.2. More detailed examples on water quality issues are provided in Chapter 7.

3.5 RECHARGE

Aquifer recharge may be accomplished by a variety of methods, ranging from surface infiltration to subsurface injection into the saturated (aquifer) zone or the unsaturated (vadose) zone. The most appropriate method of aquifer recharge depends on various factors, including the nature of the target aquifer (confined or unconfined), the quality of the source water, and land availability. Key characteristics of the primary aquifer recharge methods are listed in Table 3-1, and further details are covered in this section and in Chapters 5 through 7.

Surface infiltration is possible only for unconfined aquifers and can be accomplished through use of natural channels or with off-stream basins, pits, or trenches (Figure 3-2). Recharge of water directly to both unconfined and confined aquifers can be accomplished through the use of wells. For unconfined aquifers,

Table 3-1. Major Characteristics of Aquifer Recharge Methodologies.

Methodology	Surface Infiltration	Vadose Zone Injection	Subsurface Injection
Facility type	Recharge basins	Vadose zone wells	Recharge wells
Aquifer type	Unconfined	Unconfined	Unconfined or confined
Pretreatment requirements	Low/minimal technology	Prevention of clogging and biofouling	Prevention of clogging and biofouling
Estimated major capital costs (USD)	Low to high (depending on local land values)	Low to medium	Medium to high
Capacity	1,000–20,000 m^3/ha/day	1,000–3,000 m^3/day per well	2,000–15,000 m^3/day per well
Maintenance requirements	Drying and scraping	Drying and disinfection	Disinfection and backflushing
Estimated lifecycle	>100 years	5–20 years	25–50 years
Location of aquifer–water contact	Vadose zone and saturated zone	Vadose zone and saturated zone	Saturated zone

Source: Modified from NRC (2008).

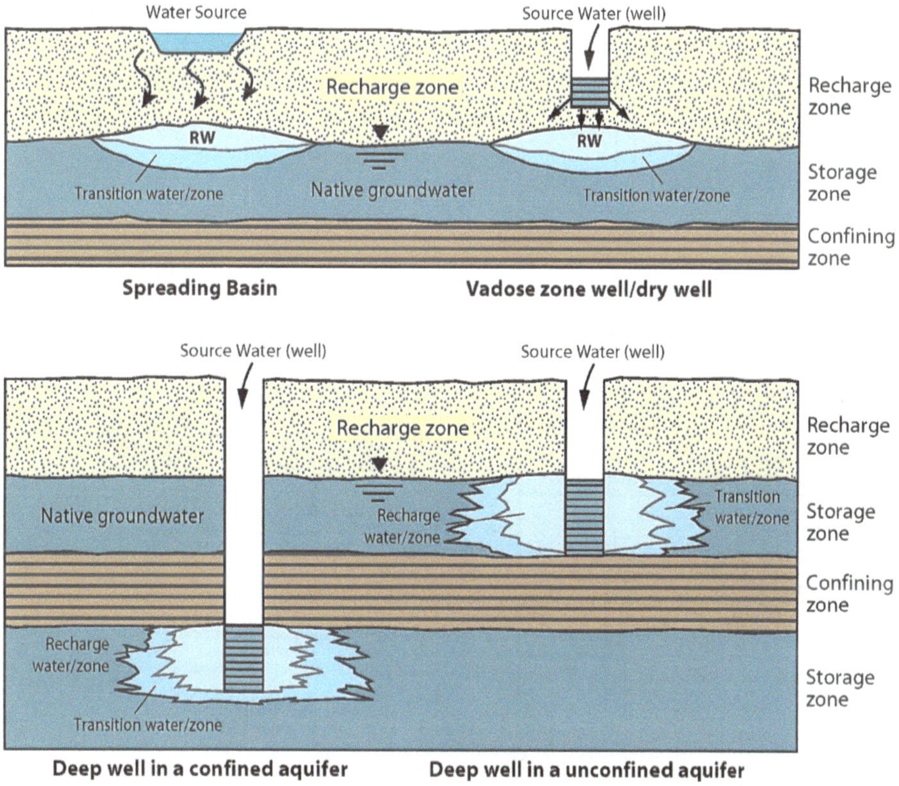

Figure 3-2. Hydrological components and processes for three recharge methods.
Note: RW = recharge water.

wells can extend into either the vadose zone (referred to as dry wells) or the saturated zone. For confined aquifers, recharge wells must extend into the saturated zone. Many excellent storage aquifers are semiconfined, with some leakage of water through overlying and/or underlying confining layers.

Figure 3-3 shows additional recharge methods. Details on surface infiltration, vadose zone, and subsurface recharge methods are provided in the following subsections.

3.5.1 Surface Infiltration. Surface infiltration methods are a viable option for aquifer recharge to unconfined aquifers if the appropriate geologic conditions exist. All surface infiltration projects require permeable materials (including gravels, sands, or loamy sands) at land surface and in the vadose (unsaturated) zone between land surface and the target aquifer. Permeable materials are necessary at land surface to allow for sufficient infiltration rates. The vadose zone should consist of sufficiently permeable soils to provide for unimpeded movement of the source water from land surface to the target aquifer. Additional considerations for surface infiltration include the following:

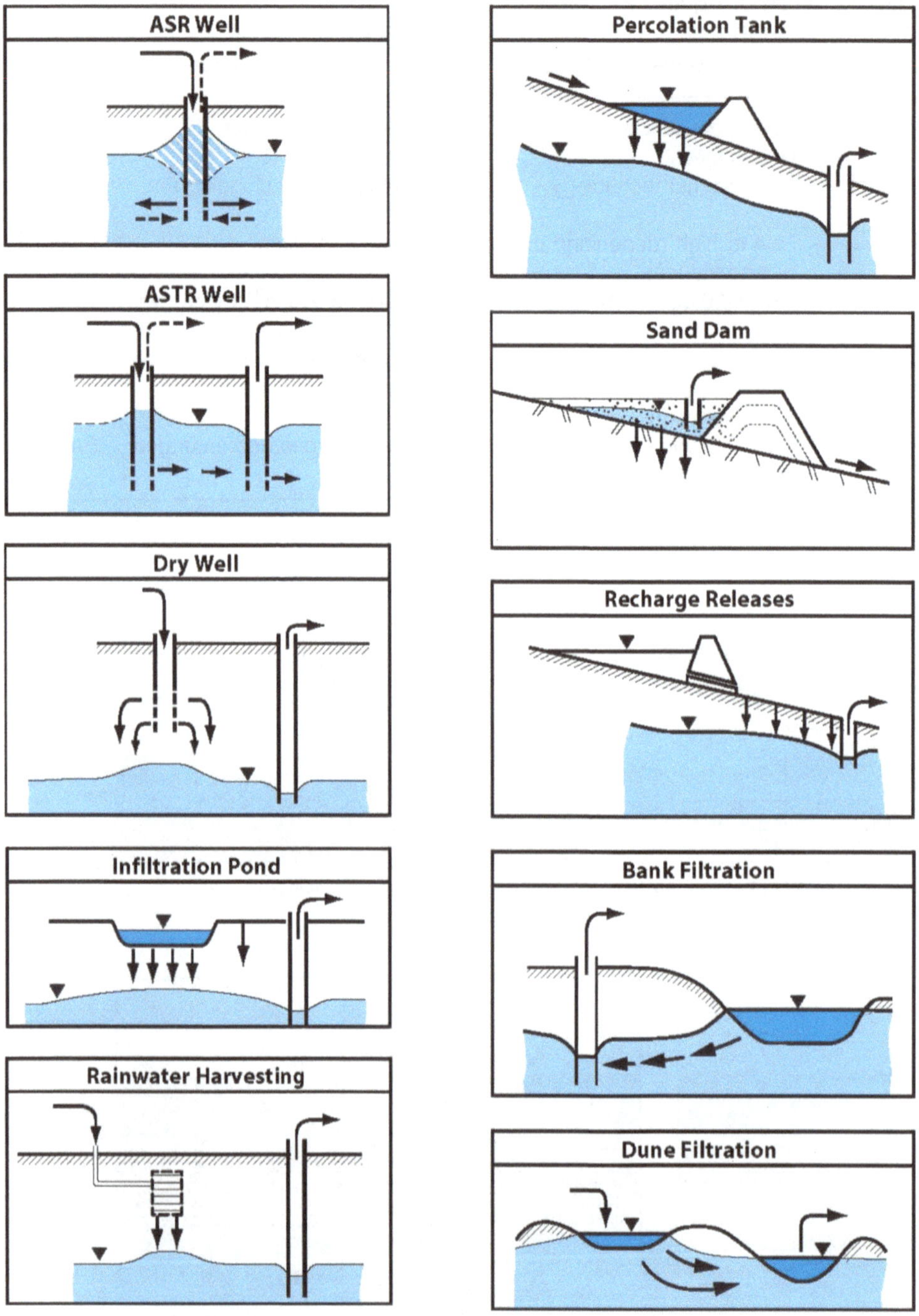

Figure 3-3. Schematic of types of managed aquifer recharge.
Source: Adapted from NRMMC et al. (2009).

- The vadose zone should be free of materials that might leach undesirable chemicals into the recharging water and not have contaminant plumes from previous spills.
- Suitable land area must be available at an acceptable cost and in adequate quantity to achieve desired recharge volumes.

Surface infiltration systems may be categorized as either in-channel or off-channel facilities. In-channel facilities range from a natural stream or river channel with no improvement to a specially prepared channel bed with automated gates and recorders. For improved in-channel facilities, recharge water may be managed by weirs, dams, levees, small dikes, or other structures that spread the water over the streambed or floodplain or increase the depth of the impounded water. The goal of these channel improvements is to increase the wetted surface area to maximize infiltration. For in-channel facilities, the surface water is moving, although sometimes very slowly. Dams must be built for in-channel facilities with adequate spillways or washout sections to handle periodic large flows. Inflatable rubber or fabric dams can be used to control lower flows but can be deflated to pass large flows. In some cases, weirs and levees are made of

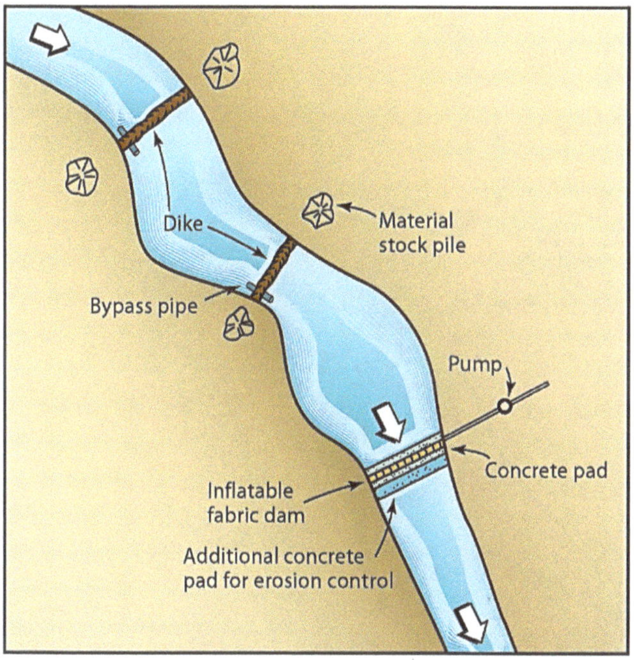

Figure 3-4. In-stream MAR system.

local soil and are considered expendable because they are easily reconstructed after damage by high flows. A schematic of an in-stream recharge system is shown in Figure 3-4.

The Bear Canyon Recharge Demonstration Project in Albuquerque, New Mexico, is an in-stream recharge system that was designed to take advantage of existing flood control structures in an ephemeral (intermittently flowing) arroyo. Additional details and photographs of the Bear Canyon Project are included in Appendix D, Case Study 1.

Off-channel facilities include ponds, basins, ditches, unused gravel pits, borrow areas to obtain soil, or similar excavations that are located away from the natural stream or river channel. In off-channel systems, the water is usually impounded and has essentially no movement. This type of system can allow suspended sediment to settle out on the bottom and sides of the excavation, which can lead to increased rates of clogging that restricts infiltration. Algal blooms that contribute to clogging also can occur in off-channel systems. Off-channel infiltration systems that receive flows from flood control channels may require large capacity intake structures to capture peak flows that may otherwise bypass the system. Off-channel infiltration basins should be hydraulically independent so that each can be flooded and then dried and cleaned by removing any fine-grained sediment that accumulates to maintain adequate infiltration rates.

The number and area of spreading basins is influenced by the magnitude and timing of high flows that would be diverted into them and by the time needed for drying of individual basins. Such drying periods should be started before infiltration rates have reached low values so that drying is achieved by infiltration only, and removal of water from clogged basins is not necessary. Low infiltration rates can occur in the winter when the water is cold, drying is slow, and infiltration recovery is incomplete, or in the summer when algae and bottom biofilms grow faster. Two types of off-stream recharge layouts are shown in Figure 3-5 for a basin-type system and in Figure 3-6 for a ditch-type system.

Relative to subsurface recharge methods, advantages of surface infiltration MAR methods include

- Low construction cost,
- Low technological requirements and ease of operation,
- Fewer requirements for pretreatment of recharged water, and
- Potential water quality benefits as water moves through the vadose zone (soil aquifer treatment, as discussed in Section 3.8).

Some of the disadvantages of surface infiltration MAR methods relative to subsurface methods include the following:

- Large land area is required,
- Indirect recharge of an aquifer may require monitoring to demonstrate that water reaches the aquifer, and
- More rapid reduction in infiltration rates over time and more frequent maintenance is needed to remove clogging material.

Additional discussion of surface infiltration system construction, operation, and maintenance is provided in Chapters 6 and 7.

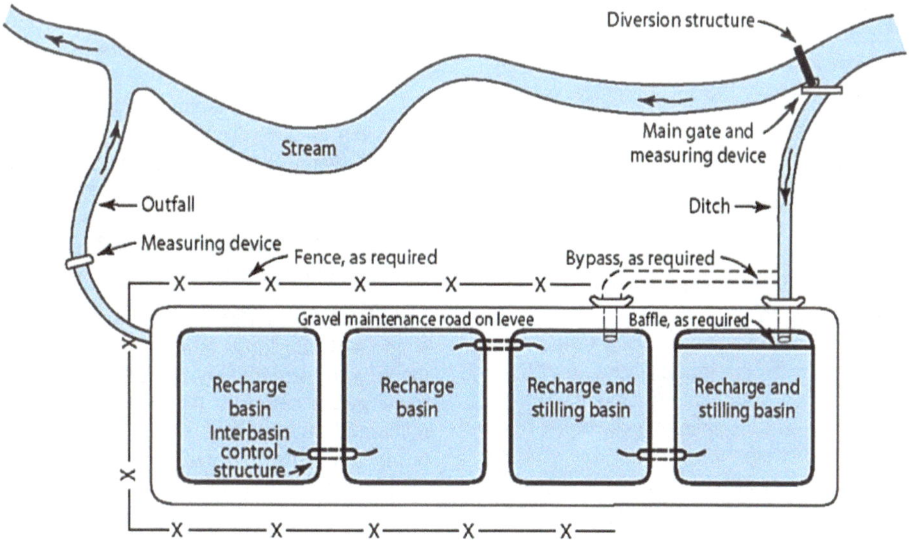

Figure 3-5. Basin-type off-stream MAR system.

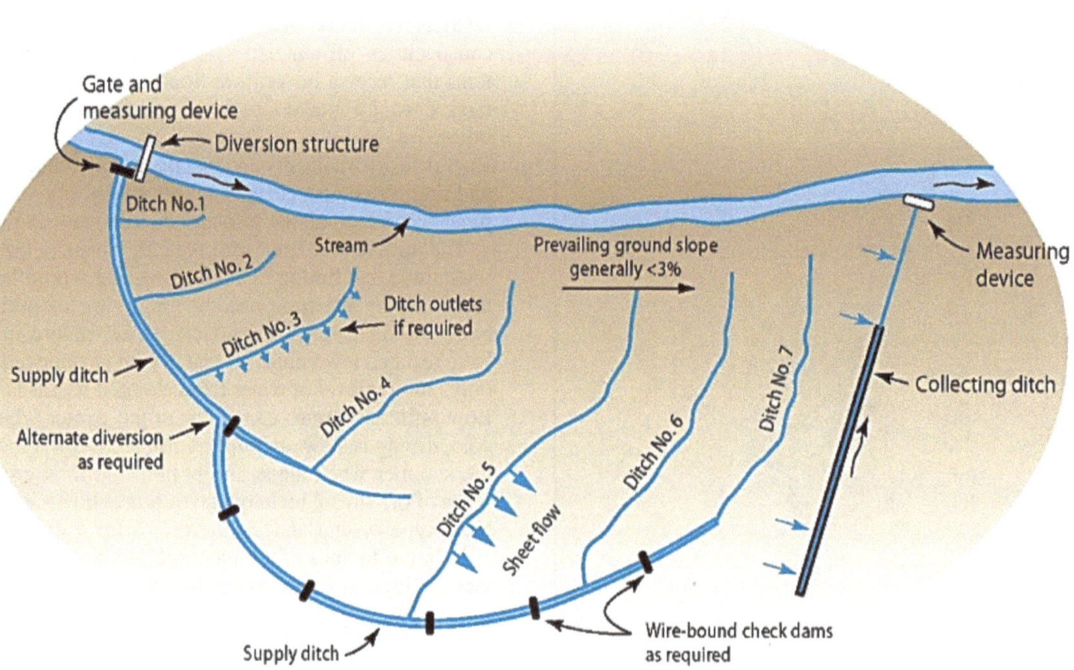

Figure 3-6. Ditch-type off-stream MAR system.

3.5.2 Vadose (Unsaturated) Zone Recharge. In cases where the appropriate conditions for surface infiltration are not available, including the presence of low-permeability zones in the shallow subsurface or site conditions not favorable for surface impoundments, water may be recharged to an unconfined aquifer by vadose zone wells or other methods that introduce water directly to the vadose zone.

Vadose zone wells are wells completed in the vadose zone above unconfined aquifers; these are also known as dry wells, because they do not intercept saturated aquifer materials. They are usually about 10 to 50 m (33 to 164 ft) deep and about 1 to 1.5 m (3 to 5 ft) in diameter, completed with a center pipe and the annular space between the pipe and the wall of the borehole filled with sand. Vadose zone wells are often used for disposal of storm runoff (Abu-Zreig et al. 2000, Bouwer 2002). Other options for recharge through the vadose zone include deep trenches, pits, adits, shafts, or similar systems that are excavated into the unsaturated zone and cover a relatively small area at the land surface.

The main feature of vadose zone recharge methods is that they do not intercept saturated materials of the local aquifer. Rather, these methods allow water to bypass the near-surface layer(s) and introduce water to the vadose zone at some distance above the aquifer water table surface. Because the recharged water still has to infiltrate through some portion of the vadose zone, some treatment benefits are provided before the water reaches the aquifer.

Advantages of vadose zone recharge methods compared with surface recharge methods include their relatively low cost, their ability to bypass shallow subsurface layers that might otherwise impede infiltration from surface infiltration facilities, and their smaller surface area. The primary disadvantage associated with these systems is their relatively short life span, which results from the difficulty of cleaning the wells to remove fine-grained materials that clog their wetted perimeter. In general, a longer life span can be achieved by recharging water of higher quality. For that reason, pretreatment methods (such as sedimentation basins) can increase the life of these MAR facilities.

3.5.3 Saturated Zone Recharge. Water may be recharged directly into saturated aquifer materials by introduction through wells. This technique can be implemented for both unconfined and confined aquifers. For purposes of MAR projects, recharge wells are generally designed to function as either dual-purpose recharge and extraction wells or just recharge wells:

- Dual-purpose recharge and extraction wells. These wells are often referred to as ASR wells. ASR wells are used for projects where recharged water is recovered from the same point where it is injected. Other than disinfection, water recovered from ASR wells usually does not require retreatment prior to potable use. Treatment is provided before recharge, thereby minimizing well clogging and reducing the frequency of backflushing.
- Recharge wells. These wells are intended to serve primarily or solely for aquifer recharge, although the well is often equipped with a pump to allow periodic backflushing to maintain injection rates. These wells are used for projects where recharged water is recovered in a different location from where it was injected or is not planned to be recovered, such as in hydraulic barriers for saltwater intrusion. Water recovered from recharge wells usually has to be treated for potable use. Long-term clogging tends to be more pronounced in recharge wells compared with ASR wells, often requiring more intensive well redevelopment procedures to maintain recharge performance.

ASR wells are designed to allow for recharge of water into a single well, for long- or short-term storage of recharged water in the area immediately surrounding the wells, and recovery of the recharged water from the same well. To allow for recovery of recharged water, ASR wells are equipped with a pump. The pump is also used to backflush the well periodically to maintain recharge efficiencies by removing accumulated clogging material, such as sediment and/or chemical precipitates.

The conversion of production wells to recharge or ASR wells is an appealing idea and has been implemented in the United

States. However, the well should be of suitable construction materials and diameter and suitable aquifer hydrogeologic and chemical characteristics. The recharge water quality must be acceptable for aquifer storage; otherwise, damage may result because of clogging in the well and surrounding aquifer formation from chemical reactions, biological growth, entrained air, and suspended particles. Conversion of production wells to recharge wells may require changes in plumbing of the well and other appurtenances.

For unconsolidated aquifers (sand and gravel), a recharge well consists of a casing, screen, gravel pack, grouting, and a pipe to apply water to the well for infiltration into the aquifer. For consolidated rock aquifers, such as sandstone, volcanic rocks, and limestone, a recharge well may be completed as an open borehole without screen or gravel pack. These wells sometimes use all or a portion of the elevation difference between ground surface and groundwater level (potentiometric surface) as the force to cause water to flow into the aquifer. However, in many cases, such as with confined or semiconfined aquifers, additional force is applied with a pump to produce a larger recharge rate. The wellhead is then designed to hold pressure.

Recharge wells can be used to inject water into an aquifer upgradient from an extraction area. Recharge wells also can be used where water-bearing formations are lenticular or discontinuous, or comprise several water-bearing strata separated by relatively impervious layers. For such formations, the screens of recharge wells can be placed opposite the water-bearing strata to be recharged. Recharge wells can also be used to create a groundwater pressure mound for effective prevention of seawater intrusion. Two types of recharge wells are shown in Figure 3-7; these consist of recharge-only or combined recharge and extraction (ASR) wells. Wells of either type can be designed for single or multiple aquifers. Equipping recharge wells and ASR wells with pumps, motors, and wellhead facilities is usually similar. Pumps for recharge wells are used for periodic backflushing to minimize well clogging. Pumps for ASR wells are used for the same purpose and are also used for recovery of the stored water.

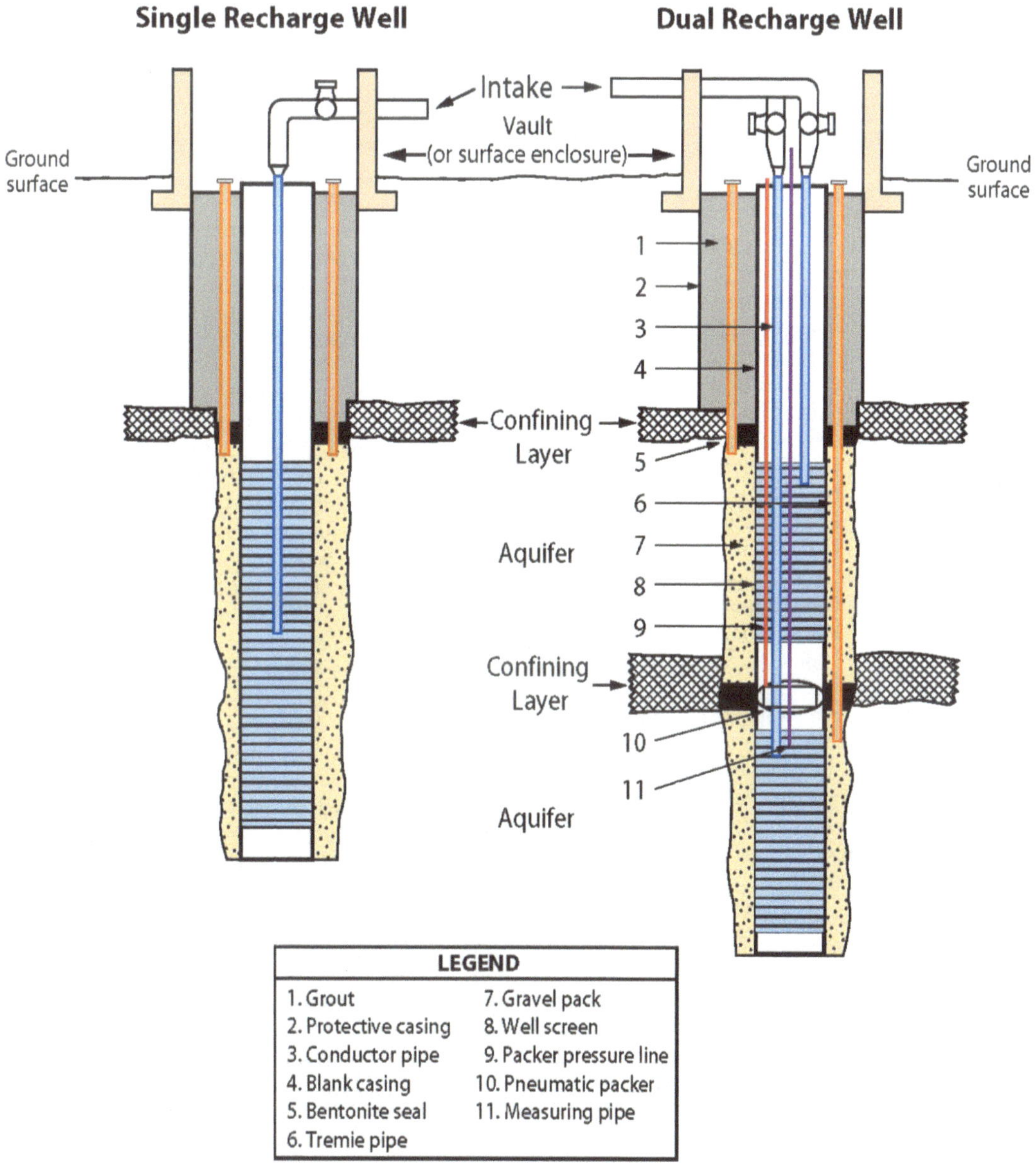

Figure 3-7. Single and dual recharge wells.

Formations containing natural openings or cavities may have high rates of recharge, causing water to flow quickly to other areas. This type of flow may occur in weathered carbonate rock (limestone or dolomite) or in fractured rock deposits, where horizontal and vertical channels, fractures, and voids can be the primary form of porosity and are the main transport mechanism for groundwater. Fault fractures may provide either a path for or a barrier to movement of groundwater. Faults may cause the separation of a basin into completely or partially isolated subbasins, may reduce groundwater flow, or may enhance the vertical movement of water between aquifers. Field investigations for aquifers located in faulted geologic settings are critical in identifying the feasibility of a MAR project.

The relatively high cost of wells, the pretreatment of the water before recharge, and required well maintenance usually make aquifer recharge by wells more expensive than recharge by surface infiltration, except in areas of high land costs or unsuitable soil conditions (Pyne 2005; NRC 1994, 2008). Additional discussion of installation, operation, and maintenance of subsurface recharge systems is given in Chapters 6 and 7.

3.6 SUBSURFACE STORAGE

Aquifer storage and retention characteristics are determined largely by the nature and extent of the aquifer material, aquifer storage coefficients and porosity, local hydraulic gradient, and the existence of fractures or faults near a potential MAR site. Generally, aquifers that underlie potential MAR facilities have extensive areas and may have experienced decreased storage (lower groundwater levels) because of groundwater extraction. In other areas, storage may be obtained by displacement of native aquifer water by the recharged water. Obtaining storage might be a goal of the MAR project when the native water may be of poor or varying quality (ranging from fresh to brackish, or high in nitrates). Currently, approximately a third of all MAR projects are in brackish and saline aquifers, displacing the ambient groundwater with stored drinking water.

The use of an aquifer as a storage basin can be related to that of a surface reservoir where portions of the storage capacity are below the spillway elevation and thus are unrecoverable, and other portions might be reserved for flood control, water supply, and other uses. These portions are referred to as the buffer zone volume and the target storage volume, respectively. The buffer zone volume is initially formed and maintained around the stored water volume. In aquifers with poor quality, the buffer zone volume is never recovered. However, once formed, it allows subsequent stored water to be recovered with very high efficiency. The usable portion of an aquifer may be further constrained by upper and lower water level requirements, by the zones penetrated by extraction wells, or by water quality problems. The amount of storage volume needed must be carefully defined and should not infringe on volumes needed by other uses or programs.

3.7 RECOVERY

Recharged water may be recovered by means of wells or by natural discharge to hydraulically connected surface water bodies. Wells can be either vertical or horizontal and can be either constructed by hand or by using mechanized equipment, such as drilling rigs. Horizontal wells are often placed under or adjacent to a stream to capture streamflow, or they may be placed approximately perpendicular to the direction of groundwater flow downgradient from recharge facilities.

In some cases, water wells are constructed in confined aquifers containing sufficient pressure to cause groundwater to flow to the ground surface (as in a flowing artesian well). MAR projects in such situations require pumping recharge water into storage, while recovery of the stored water requires reduced pumping to meet discharge pressure requirements. In most confined aquifer settings and for wells located in unconfined aquifers, pumps are usually needed to extract the groundwater and route it to places of use or storage on the surface.

MAR facilities have long been used to help maintain the flow in rivers. Water is recharged into shallow, unconfined aquifers and allowed to migrate toward and discharge into the hydraulically connected river. Alternatively, water is stored deep in ASR wells next to the river, and a portion of the stored water is used for maintenance of low flows during dry months. This storage can be a valuable water management method to increase surface water flows during seasonal dry periods, such as late summer in the Northern Hemisphere. Benefits include maintaining and enhancing riparian habitat and aquatic species. For regions in which water is managed under a priority system, an additional benefit of natural discharge to streams is that a junior appropriator may be able to implement a MAR project when demands are low, earn credit for the recharged water, and then be able to take water out of priority when demands are high. Water stored via ASR wells is less likely to be subject to regulatory restrictions to reduce groundwater withdrawals during drought periods.

3.8 POST-TREATMENT

The intended use of the recharged water is the primary reason for post-treatment for most MAR projects. Of practical importance for the viability of a MAR project are potential interactions between source water, the native water and the aquifer matrix mineralogy, and the resulting water quality and post-treatment requirements.

Underground formations act as natural physical, biological, and chemical filters that can remove undesirable constituents from lower quality recharge water as it moves through them. Improvement in the quality of the aquifer water has been the main objective of some MAR projects, using the soil and the aquifer as a treatment facility. In general, the biggest improvement in water quality occurs in the vadose zone and in the near-surface saturated zone because of their higher oxygen content and presence of naturally occurring microorganisms, which result in soil aquifer treatment (SAT; ASCE 1996; NRC 1994, 2008). For example, SAT can significantly reduce the total amount of wastewater effluent organic matter (Amy and Drewes 2007). Recharge through coarser aquifer materials or those low in organic carbon content may not produce major improvements in the quality of the water. The potential SAT of an aquifer for the source water that is intended for subsurface storage must be understood to develop adequate post-treatment processes.

For example, SAT provided efficient removal of total organic carbon (TOC) in wastewater recharged over decades at the Montebello Forebay Spreading Grounds near Los Angeles. However, when advance-treated recycled water became available for recharge, this higher quality recharge water had the potential to liberate the bound TOC from vadose zone sediments and to affect shallow groundwater. The solution was to blend various water sources to allow for more effective SAT. See Case Study 3 in Appendix D for more details.

Significant natural treatment also occurs as recharge water moves through confined and semiconfined aquifers beneath the vadose zone, including physical, microbiological, and geochemical treatment processes (Amy and Drewes 2007; Bouwer 1991;

CSIRO 2016; Lee and Jones-Lee 1995a, b; NRC 2008; NRMMC et al. 2009; Pyne 2005; Sheng 2005; Sloss et al. 1996). These processes are summarized here and are discussed in detail in Sections 7.2 and 7.3 of Chapter 7.

- Physical processes include filtration of any particulates in the recharge water or ferric hydroxide flocs (precipitates) formed underground as a result of reaction between iron-bearing minerals in the aquifer and oxygen in the recharge water.
- Microbial processes are effective at attenuating disinfection by-products in the recharge water, which include haloacetic acids and trihalomethanes. Attenuation of pathogenic viruses and bacteria also occurs during aquifer storage at rates typically expressed in days per log cycle. Storage deep underground can achieve several log cycles of pathogen attenuation.
- Geochemical processes include adsorption, desorption, solution, precipitation, cation exchange, clay swelling, and several others. These processes occurring deep underground in confined and semiconfined aquifers can mobilize and also can attenuate arsenic present in aquifer minerals. Attenuation typically requires initial formation and maintenance of a buffer zone around the well.

Until 2013, federal regulatory policy in the United States effectively precluded any allowance for natural underground treatment processes in confined and semiconfined aquifers. Recharge water had to meet all drinking water standards before recharge. A policy change by the US Environmental Protection Agency in 2013 provided greater discretion to individual states to allow time and distance for natural treatment processes underground around an ASR well or a recharge well (USEPA 2013). Demonstration of ownership or control over surrounding property beneath which such natural processes occur is necessary.

If the intended end use of stored water is for potable purposes, then additional treatment may be required before recharge or after recovery from the aquifer, or both. In most cases, pretreatment and natural treatment in the aquifer around the well is sufficient so that the only post-treatment required is to restore the disinfectant residual. For recharge wells, pretreatment would include whatever processes are needed to prevent or minimize well clogging and to obtain a recharge permit. Recovery of the stored water from another well would produce a water quality blend between the recharge water and the native groundwater. Post-treatment for potable use would typically include whatever treatment processes are needed to meet potable standards, or to meet standards applicable for the intended use.

Groundwater below septic tank leach fields typically contains elevated levels of total dissolved solids (TDS), nitrite, nitrate, chloride, bacteria, and viruses (Bouwer 1978). Groundwater below irrigated fields usually contains higher levels of TDS, nitrate, pesticide residues (Bouwer 1991, Yadav et al. 2002), and in some areas, trace elements that were leached from the soil and unsaturated zone (including selenium, boron, arsenic, molybdenum, cadmium, mercury, and others) (Khan et al. 2007, Murtaza et al. 2008). These settings may require additional postrecovery treatment before end use.

3.9 END USES

Recovered water can be used for drinking water supplies, industrial uses, agricultural irrigation, and maintenance of riparian habitat and aquatic ecosystems (NRMMC et al. 2009, NRC 2008). To date, 29 different applications and end uses of stored water have been identified for ASR wells (Pyne 2005). Selection and prioritization of these goals is an important initial step in project planning. Some of the more common applications are seasonal storage, long-term storage (or water banking), emergency storage, achieving water supply redundancy and reliability, maintaining pressures and flows at distal portions of a water distribution system, and deferring expansion of water treatment plants and transmission pipelines by greater reliance upon aquifer storage.

The intended use of recovered water is the most important factor that determines the economics of MAR facilities, along with many of the decisions regarding site selection, aquifer selection, recharge and recovery methods, timing and duration of storage, available operations of source waters, requirements of pre- and post-treatments, and permitting and regulatory constraints. The final use of the stored water also dictates the desired water quality (NRC 2008).

CHAPTER 4
PLANNING

Planning is the first of five phases that should be undertaken when developing and implementing a MAR project. The five phases of a MAR project are depicted in Figure 4-1. The planning process defines project objectives and helps identify the potential sites and MAR methods that should best meet those objectives. Data collection and analysis activities, discussed in Chapter 9, are then conducted on the candidate site or sites to assist in design, the second MAR phase, discussed in Chapter 5. Construction and operation of a MAR project constitute the third and fourth phases and are discussed in Chapters 6 and 7, respectively. The final phase of a MAR project is closure, which is discussed in Chapter 8.

Planning is a sequential process that starts with defining project objectives, gathering available data on the potential project site or sites, and then evaluating those data against criteria that are developed to assess the suitability of the site(s) for MAR activity. Figure 4-1 outlines the planning process within the context of a MAR project. The data gathering and site evaluation phases are most effective when conducted in an iterative process, with increasing levels of data and evaluation as potential sites continue to appear favorable for MAR project implementation. The probability of successfully implementing a MAR project can be increased by following a systematic process as described in this chapter.

This chapter starts with initial scoping of a project, which includes defining MAR objectives and developing a preliminary conceptual plan (Section 4.1). Next, Section 4.2 discusses the types of preliminary data that might be available to identify and prioritize candidate MAR sites and recharge methods. Often there are several potential sites and recharge methods that could meet the objectives of the MAR project. Possible methods to evaluate alternatives, including regulatory and institutional issues, are also presented in Section 4.2. Section 4.3 discusses

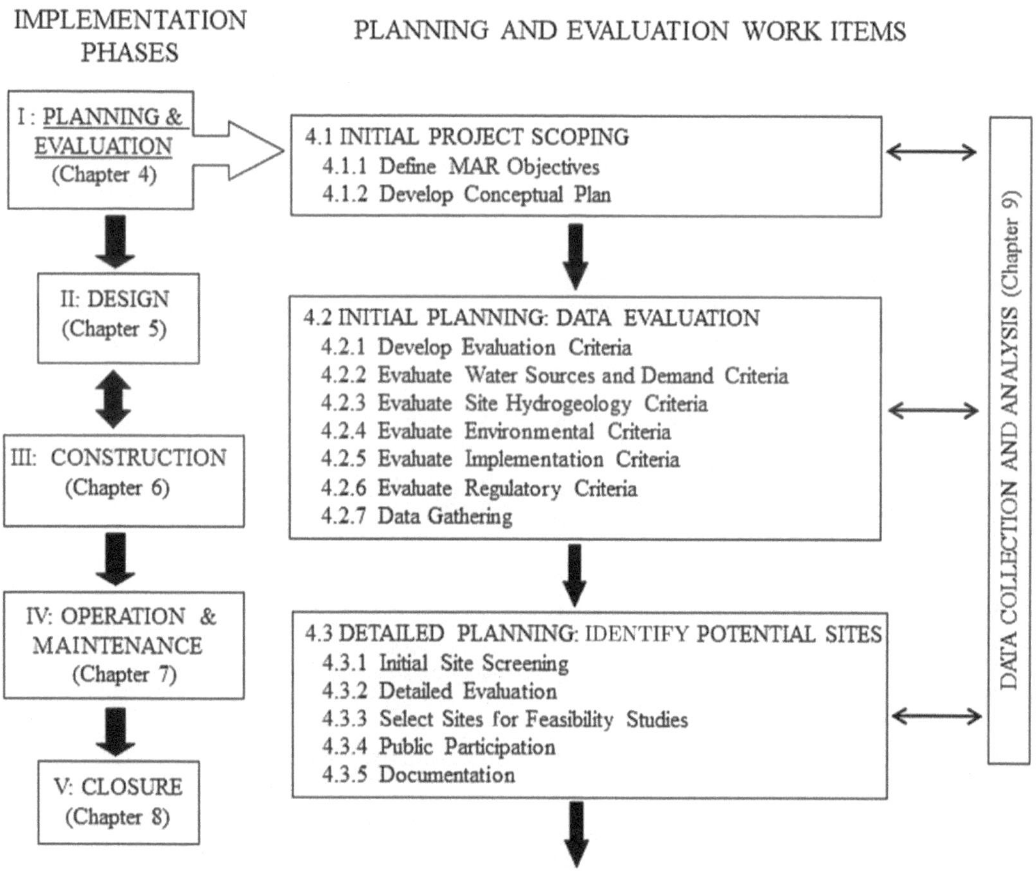

Figure 4-1. Details of the planning and evaluation phase leading to design, construction, and operation of a MAR project.

recommended steps to implement a detailed evaluation of potential MAR project sites, including public involvement and documenting the outcome of the planning process.

4.1 INITIAL PROJECT SCOPING

Complex projects such as MAR projects are best implemented both when there is a clearly defined list of project objectives and when a plan for the MAR project has been developed at a preliminary or conceptual level at the start of the project. These activities are considered initial project scoping tasks within this standard. Thoughtful attention paid to each aspect of the scoping process serves projects well by helping to focus on the much more time-consuming data evaluation, site assessment, and design activities that follow.

4.1.1 Define MAR Objectives. The objectives for any MAR project should be established early in the planning process. The objectives should be the basis upon which candidate MAR sites and recharge methods will be evaluated. Project objectives are usually developed by the technical project team that is undertaking a MAR project. If there is a potential for the MAR project to affect the public or other nonproject entities, then it may be beneficial to include those entities when developing objectives or have them review the objectives as a means to gauge stakeholder understanding and support for the MAR project.

Project objectives also help guide the type of information that should be collected during the planning process. For example, if an objective of a MAR project is to store water for some future use, then the amount of water needed by that future use should be thoroughly researched and well defined. It is common for there to be multiple objectives for a MAR project that include water supply, water quality, and environmental considerations. In such cases, the multiple objectives could be categorized by relative importance or could be classified by using a ranking system, such as primary and secondary objectives, to help when evaluating multiple potential sites and recharge methods.

Developing a list of several objectives can broaden the base of support for the MAR project while also helping to refine its conceptual design. For example, a community may have a primary objective of storing potable water underground to meet peak water demands. Secondary objectives could be to delay the need to secure additional supplies and to delay the need to expand water treatment facilities. Another example would be for a supplemental water supply that could serve areas with low groundwater levels in peak demand months. Multiple objectives can be met in this case by locating MAR recharge facilities and recovery wells at the subject area because excess water would be stored there during low demand months and would increase water levels. During peak demand months, the water would be recovered, helping to maintain adequate groundwater levels. An example of multiple-objective surface recharge facilities is one that provides both flood control and recreation benefits. In this example, flood detention basins can be constructed along with settling basins to provide a source of recharge water, and the basins could be surrounded by recreation areas, which may also provide environmental benefits, such as habitat.

The Tamarack Ranch Recharge Project in Colorado is an example of a multipurpose facility managed by three states (Colorado, Wyoming, and Nebraska) and the US Department of the Interior (see Appendix D, Case Study 4). The program involves pumping 17 wells adjacent to the South Platte River during the winter low-demand period, conveying the water to surface recharge in 11 ponds (totaling 32 hectares (80 acres), and timing the accretions back to the stream in the summer to augment flow to benefit multiple endangered species. Colorado water users can also lease excess accretion credits for private augmentation plans.

Possible objectives for a MAR project, derived from projects currently in operation or in various stages of development around the world, include many of the following (Topper et al. 2004):

- Manage water supply:
  - Provide seasonal to long-term storage,
  - Improve consistency of supplies,
  - Improve security of supplies,
  - Provide strategic emergency supplies,
  - Enhance well field production,
  - Defer expansion of water facilities,
  - Improve flood control,
  - Provide means for discharge of treated wastewater, and
  - Maintain pressures and flows in water distribution systems,
- Manage water quality:
  - Improve surface water quality,
  - Improve groundwater quality,
  - Reduce disinfection by-products, and
  - Provide additional water treatment,
- Restore the aquifer:
  - Restore water levels to former higher levels,
  - Reduce subsidence, and
  - Prevent or reduce seawater and brackish water intrusion,
- Protect the environment:
  - Maintain wetlands,
  - Increase streamflow through increased groundwater discharge,
  - Maintain minimum flows and levels through direct discharge during droughts,
  - Enhance riparian habitat,
  - Provide more uniform surface water temperature,
  - Control contaminant plumes, and
  - Reduce greenhouse gas emissions from groundwater pumping,
- Meet legal obligations:
  - Offset out-of-priority stream depletions caused by pumping, and
- Other:
  - Provide thermal energy storage.

4.1.2 Preliminary Conceptual Plan. Once MAR objectives have been defined and prioritized, a conceptual plan should be developed. At this stage in the planning process, the conceptual plan should attempt to describe, in general terms, recharge sites and methods that would best meet the project objectives. This preliminary conceptual plan forms the framework for gathering available data to evaluate potential sites and recharge methods (Section 4.2) and is the basis for developing criteria and scoring factors to screen potential sites and aquifers or storage intervals within aquifers. As with data gathering and evaluation, conceptual plans should be developed in an iterative manner, with increasing detail provided as more data become available and the number of potential MAR alternatives is reduced. The conceptual plan also becomes a key element when the planning process and results are documented, as discussed in Section 4.3.5.

4.2 INITIAL PLANNING ACTIVITIES

An effective method to assess potential MAR sites and recharge methods at the planning stage consists of developing evaluation criteria, gathering available data for each of the criteria, and then undertaking a ranking or scoring process to identify the sites and recharge methods that best meet the criteria. Potential criteria and the data that might be available for this level of analysis are discussed in this section. Methods to rank or score potential sites and recharge methods based on the criteria are discussed in Section 4.3.

4.2.1 Develop Evaluation Criteria. A series of criteria should be developed to evaluate the feasibility of implementing a MAR project operation based on the objectives that have been established for the project. Using evaluation criteria is particularly useful to help prioritize alternative sites and recharge methods within the study region. The criteria should take into account a series of factors that are likely to include water supply and demand and hydrogeologic, environmental, and implementation factors. Possible criteria are listed in Table 4-1 and are described in detail further in this section. The amount of detail needed to define each criterion adequately varies depending on the project location, size, available information, and the anticipated MAR recharge method. In general, the criteria should be quantitative, as discussed later, but it is still useful in the MAR evaluation process to develop qualitative criteria that can be used in the site scoring process. Some of the criteria may need to be reviewed and revisited in more detail in the later design and construction phases of the project.

The criteria developed for a project should relate directly to the MAR project objectives. In many cases, a given criterion, such as adequate aquifer storage, might relate to several project objectives. Care should be taken to define enough criteria that at least one relates to each project objective. Additional criteria may be needed to satisfy the public or other concerns.

Potential criteria are discussed here and are modified from recent regional MAR assessments (CWCB 2007, OWRB 2010). Additional criteria may be more relevant for a given site, but the ones described here are commonly used and help address typical objectives of a MAR project.

4.2.2 Example Water Sources and Demand Criteria. Careful consideration of quantity and quality of alternative sources of water and the receiving groundwater, as well as their compatibility, is essential to a successful project. It is common, especially in arid and semiarid regions, for potential MAR project sites to have suitable hydrogeologic characteristics but to be limited either by lack of water for recharge or by poor-quality water that is economically prohibitive to treat before recharge.

Availability of Water. Potential water sources for recharge include surface water from streams, lakes, or reservoirs, reclaimed wastewater, stormwater runoff, groundwater from other aquifers or from the same aquifer at a different location, and treated drinking water. Surface water sources are usually the least expensive and most commonly used sources for recharge. Each potential water source should be evaluated considering the average flow available; variability (daily to very long periods) in both flow rate and quality; and trends in flow and quality based on hydrology, competing demands, legal or regulatory constraints, or other characteristics of each system. Water availability typically declines during the growing season as higher priority demands for that source arise. Conversely,

Table 4-1. MAR Evaluation Criteria.

Category	Criterion	Description/Applicability
Water sources and demand	Availability of water	Whether water is available physically or institutionally; trends and variability
	Proximity of source water	Distance from water source to MAR site
	Source water quality	Quality of source water untreated or post-treated
	Estimates of water demand	Trends and variability in demand from MAR site under anticipated operation scenarios
Site hydrogeology	Hydrogeologic suitability	Whether aquifer characteristics will allow anticipated recharge
	Amount of available storage	Physical space in aquifer, unsaturated pore space for unconfined aquifers, pressure head and porosity for confined aquifers
	Residence time	How long water will stay under dominion and control for aquifer setting
	Induced seismicity	Likelihood of causing a seismic event
Environmental considerations	Waterlogging and nonbeneficial use	Where elevated water table conditions may affect soils and structures or be lost through evapotranspiration
	Habitat concerns	Possible effect on sensitive environments
	Effects on aquifer water quality	Effects of introducing water with differing chemistry
Implementation considerations	Proximity to existing infrastructure	Affects overall cost
	Proximity to demand	Affects overall cost
	Land ownership and use	Affects cost and permitting
	Cost	Total cost to implement and maintain
	Site access and security	Affects cost and permitting; applies to protection of water supply
	Conditions surrounding site	Affect costs, permitting, and environmental considerations
Regulatory considerations	Permitting and other regulatory requirements	Affect cost and permitting

there often is excess water available during the winter months when demands are low.

The availability of source water for a MAR project is often the limiting factor for a technically efficient and cost-effective level of operation over the long term. If possible, alternate sources of water should be identified to supplement the primary source if it becomes unavailable. The degree of availability of a potential water supply should be evaluated by using records of historical flows and, potentially, taking into account projected effects of climate change on catchment hydrology. Adverse effects of interrupting MAR project operations because of lack of supply will vary based on the objectives of the MAR activity, the main beneficiary (such as agriculture, residential, commercial, or environmental), and the timing and amount of shortage. Typically, MAR projects can operate successfully even with interruptible supplies, operating when water is available.

Proximity of Source Water. Proximity describes the distance between available sources of water to be used for recharge and the MAR project facilities. At the planning stage, this criterion can be characterized on a qualitative basis and relates primarily to the costs of water conveyance because shorter conveyance distances are more desirable. In many cases, the streams and existing pipelines, canals, and ditches could be assumed to be available to deliver source water closer to the MAR facility, so this criterion considers the additional conveyance systems needed. This criterion is also used to consider whether pumping or lift stations would be needed in cases where the MAR facility is located at a higher elevation. In such cases, gravity feed conveyance systems are more desirable.

Source Water Quality. The quality of water being recharged, including its physical, chemical, and microbiological characteristics, should be evaluated to assess its potential for causing undesirable effects in the recharge facilities and in the target aquifer. Various kinds of chemical reactions can occur and could affect the hydraulic performance (storage and transmission) of the aquifer. Of particular concern is the potential for chemical reactions that create insoluble species that plug the aquifer, such as the formation of calcium carbonate, ferric hydroxide, and several manganese oxide precipitates. Other reactions, such as cation exchange, dissolution, and the expansion of clays, can occur in an aquifer if the source is chemically incompatible and can significantly change the hydraulic properties of the aquifer. In addition, nutrient- and organic-rich recharge waters can promote biological growths that can plug well screens and aquifers. These effects can be mitigated by treating the water both before recharge and/or after extraction of recharged water and by more frequent maintenance of recharge facilities, including well backflushing and rehabilitation and scraping of recharge basins. Pre- and post-treatment would lead to increased implementation costs and are therefore less desirable.

Available water quality data should be reviewed during the planning phase for general quality characteristics and for any seasonal or long-term trends in quality that can affect recharge activities. For example, months when high surface water flows are available for recharge can also be months when increased turbidity occurs that could increase the likelihood for recharge basins or recharge wells to plug without adequate water pretreatment. Other factors that should be considered include the presence of nutrients, high concentrations of effluent, high salinity, or total dissolved solids (TDS). A thorough evaluation of recharge water quality for each source is required before or during the design phase to assess the potential for problems and treatment requirements properly.

Estimates of Water Demand. Consideration of the magnitude and variability in water demand indicates the potential volume of water that might be available for recharge and needed for recovery. In many situations, it may be useful to estimate average monthly demand not provided by other sources to justify adding MAR as a potential component of the water supply portfolio. In addition, it can be helpful to compute ratios of projected maximum daily to average annual demand, maximum weekly to average annual demand, and monthly demand as a percent of average annual demand, in assessing the duration of peak demand periods when recovery of recharged water from the potential MAR project would provide maximum benefit. These ratios can be used to assess the amount of idle existing supply, treatment, and transmission capacity available during each month. Analysis of daily demand records usually identifies opportunities for storage that may be missed in monthly analyses, leading to substantial savings in storage volume and associated costs.

4.2.3 Example Site Hydrogeology Criteria. Hydrogeologic studies are usually the most critical element in ranking potential MAR sites. Careful evaluation of area hydrology and geology can lead to the selection of suitable underground storage zones, water treatment requirements, and the location and type of recharge facilities. It is frequently possible to complete this analysis during the planning phase without additional field investigations, although site-specific data generally are needed before final design.

The general hydrogeologic evaluation of a potential MAR site should consider the following technical issues. This information can be used to assess the hydrogeologic suitability of recharging and extracting water from target aquifers, to estimate the amount of available underground storage, and to characterize potential geochemical reactions within the host aquifer.

Hydrogeologic Suitability. Hydrogeologic suitability refers to the ease with which project water can infiltrate into the ground and percolate to the water table or can be injected into a targeted aquifer. A higher rate is more favorable. Suitable hydraulic properties of the target vadose zone or aquifer are important to maximize recharge of project water, which is often available on a limited basis, such as during high streamflow associated with heavy runoff events. Even if recharge water is available on a consistent basis, this criterion is important for efficiency of a recharge project and for ease in extracting the recharged water at a later date.

For recharge methods involving surface or vadose zone infiltration, the hydrogeologic suitability should be evaluated through a characterization of the surface soils and underlying vadose zone. In many arid regions, the depth to groundwater is significant, so infiltration characteristics of the vadose zone become an important consideration, particularly if low-permeability layers are present that might limit infiltrating recharge water from reaching the water table. Soil maps and soils data are widely available in the United States through the Natural Resources Conservation Service (NRCS), an agency of the US Department of Agriculture, and other agencies and can be used to estimate shallow infiltration rates. Soil maps are useful for planning purposes, and site-specific data can be confirmed later during field investigations if a site is deemed suitable for further evaluation. Vadose zone characteristics, including lithology, texture, and hydraulic properties can be inferred from geologic logs in nearby wells, if available. Regional vadose zone data may also be available from groundwater basin studies, environmental investigations, stream infiltration studies, or studies of natural recharge. Site-specific conditions need to be more fully characterized using field investigation techniques, including percolation testing.

Hydrogeologic suitability beneath a potential MAR site can be estimated from the hydraulic conductivity (K) or transmissivity (T) of the aquifer and, for unconfined aquifer settings, the infiltration capacity of the overlying unsaturated soil. These hydraulic properties can be estimated either by field testing or in the laboratory. In early stages of an investigation, the hydraulic properties may only need to be estimated using values available in the groundwater literature that relate hydraulic conductivity to aquifer sediments or rock type (for example, Fetter 2001, Freeze and Cherry 1979, Heath 1984). In many areas in the United States, there are existing studies conducted by the US Geological Survey or other agencies that may provide information sufficient for characterizing hydrogeologic suitability at the planning phase. The US Geological Survey currently maintains the National Water Information System (NWIS) database (https://waterdata.usgs.gov/nwis), for example, which contains water level and hydraulic property data for many groundwater systems in the United States. Other sources of hydrogeologic information include specific capacity tests included in some well permits, which can be used to estimate aquifer transmissivity (T) values.

Another consideration is the potential for geochemical interactions between the aquifer matrix and recharge water to affect the hydraulic properties of the target aquifer or vadose zone. As an example, dissolution of subsurface minerals could lead to precipitation and clogging of the target intervals.

Detailed hydrogeologic characterizations were conducted for both the Bear Canyon Recharge Project and the Rio Rancho Direct Injection Demonstration Projects (Figure 4-2 and Case Studies 1 and 10 in Appendix D). Both of these projects included extensive hydrogeologic characterization, in part because they were the first two recharge projects in the state of New Mexico and, thus, subjected to greater scrutiny. Appendix C summarizes MAR regulations for selected states.

Amount of Available Storage. This criterion describes the volume of storage that is available in the subsurface. The available storage volume can be estimated using the thickness of the zone available for recharge, the storage capacity of the soil or rock that will receive recharge water, and the area of influence of MAR project operations.

The thickness of the recharge zone needs to be estimated differently for confined and unconfined aquifers. For an unconfined aquifer, the thickness represents the distance between the water table and the land surface and refers to the unsaturated zone above the aquifer. For confined aquifers, the thickness represents the ability to inject water under pressure into a confined aquifer, which is partly dependent on the depth to the potentiometric surface below land surface and also the pressure that may be applied at land surface during recharge periods. Typically, this applied pressure should be less than the drawdown available by pumping the well during periodic backflushing events. For unconfined aquifers, it is often appropriate to exclude from the volume calculations a certain thickness extending below ground surface to minimize potential problems associated with water levels rising too close to the land surface, including ponding, declining infiltration rates, flooding of basements, waterlogging soils, and/or nonbeneficial use by deep-rooted plants that draw water from the saturated zone (phreatophytes). A reasonable interval for these purposes is to plan for water levels to rise no more than to within 3 to 5 m (10 to 20 ft) of land surface. However, this is a site- or project-specific issue that should be decided by the planning team. Trends and fluctuations in vadose zone thickness over time need to be considered. Seasonal or long-term changes in the water table caused by groundwater use, natural recharge, or other factors may influence the available storage capacity of the MAR project. The thickness calculations may be limited by lack of site-specific knowledge on the depth to water or depth to the potentiometric surface.

The storage capacity of the sediments receiving recharge water is defined by the specific yield for an unconfined aquifer and by the porosity and storage coefficient for a confined aquifer. For materials with good storage potential, typical values for specific yield are 10% to 30% (0.1 to 0.3), and typical storage coefficient values range from 0.001 to 0.00001 (Heath 1989). Porosity for confined aquifers typically ranges up to 50% (Fetter 2001). These terms are expressed in units of volume of water per unit volume of aquifer material. Values are derived from field testing or can be estimated using literature values. Natural variations in the composition of sediments or rock can result in a range of values for specific yield, porosity, and storage coefficient, even at a site scale, so average values representative of most of the permeable subsurface materials should be chosen.

The appropriate storage term (specific yield or storage coefficient and porosity) is multiplied by a unit of length (the available aquifer thickness in the case of an unconfined or confined aquifer; the anticipated change in hydraulic head in the case of a confined aquifer) to compute a maximum potential storage volume per unit surface area, such as cubic meters per square meter or acre-foot per acre. This is often the most useful way to

Figure 4-2. Surface recharge at Bear Canyon recharge demonstration project.

express the maximum amount of available storage when evaluating multiple potential sites for MAR project operations. However, this storage volume may be limited by well recharge rates or other factors.

The available storage volume in the confined portions of aquifers is a measure of the available potentiometric surface. This storage volume occurs because the confined aquifers are completely saturated with water and changes in storage are governed by the compressibility of the aquifer skeletal matrix and compressibility of water. The potentiometric surface of a confined aquifer represents the height above a common reference frame (for example, altitude relative to mean sea level) that the water level would rise in a sealed well open to the atmosphere and screened only in the confined aquifer. From a practical viewpoint, the available storage volume in the confined portion of aquifers usually far exceeds water storage volume requirements. The native groundwater is displaced around the well and replaced by the usually higher quality water to be stored and ultimately recovered.

The available storage for the MAR project can be limited by the ability of the subsurface sediments to transmit the water away from the site and the acceptable area over which water levels can rise. The acceptable area of mounding can be affected by various factors, including adjacent land uses and geotechnical considerations, such as increased hazard of liquefaction in earthquake-prone areas. The available storage should also take into consideration that a buffer zone is created at the outer edges of the recharged water. The buffer zone separates the recharged water from the groundwater in the aquifer and is usually not recovered. The volume of water in the buffer zone reduces the volume of usable water calculated beneath the project site, with the reduction depending on the quality of the extracted water and its intended uses.

Residence Time. This criterion considers the amount of time a recharged volume of water remains in the aquifer before it is recovered or discharges naturally without the use of additional infrastructure. The residence time for water in unconfined aquifers is proportional to the distance between the MAR facility and a point of discharge, such as a nearby stream or lake, and could be as short as days to weeks. Conversely, the residence time for water in confined aquifers (those that are not dominated by fracture flow or carbonate dissolution conditions) is anticipated to be much longer (years to decades or longer). This delay occurs because the water in confined aquifers usually migrates very slowly, though some systems may be hydraulically connected to overlying alluvial aquifers or adjacent surface water bodies. Residence time usually refers to the natural flow processes within an aquifer and does not address the use of engineered structures such as slurry walls that would increase residence time because this would be a site-specific implementation issue.

Typically, there is a desire to have access to the recharged water for one or more years, so a long residence time is more desirable. However, depending on the objectives for the MAR system, residence times less than several months may be desirable. Well and well field design can often address potential losses caused by lateral flow in the storage aquifer, such as by orienting wells in an upgradient–downgradient direction, recharging into upgradient wells, and recovering from downgradient wells.

There are a variety of methods to estimate residence time, ranging from simple calculations based on Darcy's law to more complex analytical and numerical flow modeling solutions that consider the decay of recharge mounds over time, to detailed field monitoring. Chapter 9 discusses example methods.

Induced Seismicity. In areas known to be seismically active, consideration should be given to the possibility of induced seismicity from increases in pore pressure within deep aquifers, particularly for sites using deep ASR wells. The direct relationship between induced seismicity and the injection of wastewater into deep bedrock aquifers is well documented (NRC 2013). The relationship has received considerable attention since the 1990s with the increase in deep disposal of wastewater from oil and gas well stimulation and production. A similar relationship has not yet been documented for the injection of freshwater for storage. This lack of documented impacts probably reflects the low injection pressures typically used for ASR wells, normally below about 138 kPa (20 pounds per square inch [psi]) at the wellhead, compared with much higher injection pressures for deep wells used for wastewater disposal and hydraulic fracturing, which can be up to 7,000 kPa (1,000 psi). Because of increased concern from the public as well as federal and local regulatory agencies, the potential for induced seismicity from deep aquifer storage should be considered in the early planning stages of a MAR project, particularly in areas of known high seismic hazard risk. The issue may also arise where MAR is being considered for frontier areas, where there has not already been depletion of groundwater from deep aquifers.

Induced seismicity from deep fluid injection occurs in areas where natural tectonic stresses are favorably oriented relative to natural existing fractures and of sufficient magnitude such that even slight increases in fluid pressure can allow movement along the existing fractures. In many cases, the fractures may be faults exhibiting previous displacement from the geologic past. Shallow earthquakes result from slip along existing fractures when the effective stress acting along the fracture exceeds frictional resistance to slip. An increase in fluid pressure within the fracture can overcome the force of friction relative to the effective stress, allowing slip on the plane. The increase in pressure can be much less than required to actually fracture the rock.

Additional geologic data may be necessary to evaluate the potential for induced seismicity after initial screening of a potential MAR site indicates that seismic hazards are elevated. Chapter 9 discusses additional data requirements.

4.2.4 Example Environmental Criteria. Several criteria relating to potential environmental and ecological effects from a potential MAR project should be considered during the planning process. These effects could have a bearing on the permitting process, which is described in Section 4.2.6.

Waterlogging and Nonbeneficial Use. This criterion considers the potential for shallow water table conditions, both near a potential recharge area and downgradient of the area, which could promote the growth of undesirable vegetation such as tamarisk (also known as salt cedar) and lead to nonbeneficial water consumption. The shallow water table conditions could also lead to waterlogging of soils (including increasing liquefaction potential), creation of undesirable wetlands, and flooding of nearby basements. The potential for these concerns can be evaluated using the depth to water table maps and estimates of mounding caused by anticipated recharge volumes. The maximum depth to which this criterion is likely to have an impact is approximately 10 m (33 ft) below ground surface, which is the approximate depth to which some nonbeneficial species, such as tamarisk, are capable of extending their roots. This criterion applies mainly to alluvial systems and unconfined aquifer settings. Underground water storage in the confined aquifers is expected to have minimal effects regarding this criterion.

Habitat Concerns. This criterion addresses the presence of habitat for federally designated threatened and endangered (T&E) species, as well as wetlands, that could be adversely impacted by construction or operation of potential underground water storage projects. Maps showing proximity of various T&E species habitat to recharge sites should be used in evaluating the likelihood of potential negative impacts caused by recharge activities. The US Bureau of Land Management has prepared one such map for lands under its jurisdiction titled "Areas of Critical Environmental Concern," which identifies lands as having important riparian corridors, threatened and endangered species habitat, cultural and archeological resources, and unique scenic landscapes that the agency believes need special management attention (USBLM 2006).

Depending on timing and operation of a recharge project, the impacts on habitat can be both positive and negative. For example, the recharge ponds can provide beneficial waterfowl habitat in the fall and winter months. This criterion is expected to play a minor role in the confined aquifer settings because of the small surface area associated with recharge wells and associated piping.

It may well be that an objective of a recharge project is to maintain or enhance habitat. In this case, the planning process and development of criteria must include relevant managing agencies and appropriate stakeholders.

Effects on Aquifer Water Quality. This criterion addresses water quality within the aquifer and how it affects the quality of the recharged water. It considers the potential added treatment costs when extracting originally high-quality recharge water that has been placed into an aquifer with lower water quality and the need for additional treatment when the water is extracted, assuming recovery at a separate location from the point of recharge. However, the possibility also exists that water quality will be improved in the aquifers receiving recharge water. The criterion also addresses the concern of degrading the aquifer water quality by leaching of minerals naturally found in the subsurface materials when recharge water is added. Locations where there is a high leaching potential for arsenic or other metals could necessitate well design and operational measures that have been shown to control the mobilization and attenuation of these metals effectively. Maps or data of total dissolved solids (TDS) in the groundwater can be used as an indicator for general water quality in the aquifers. In addition, this criterion should be used to consider the presence of contamination in the soil or groundwater in the vicinity of the candidate MAR site and the potential effects of recharge on contaminant migration.

4.2.5 Example Implementation Criteria. The following criteria pertain to facility siting issues that are used in addition to hydrogeologic and environmental considerations. They are related to the relative ease or difficulty of undertaking a MAR operation within a potential site or region under consideration.

Proximity to Existing Infrastructure. The presence of nearby water conveyance structures and other infrastructure can be an important consideration affecting the cost and overall feasibility of a MAR project. The proximity of ditches or pipelines that could be used to convey water from a stream or other water source to a MAR facility improves the suitability of that potential site for many reasons. These reasons include reducing construction costs and environmental disturbance, the latter of which would also help expedite permitting. Maps from state nature resources agencies, the National Hydrography Dataset (USGS 2006), and various municipalities can be used to identify key conveyance structures. If the conveyance structure is owned or operated by other entities, then agreements would need to be arranged for its use. In many cases, water is available for recharge on a seasonal basis during low water use times. At these times, there could be excess capacity in the infrastructure, which would facilitate its use for recharge purposes.

Proximity to Demand. The distance between a potential MAR project location and areas of water demand can affect the cost of MAR project operations and thus the suitability of one site over another. Projects are more likely to be implemented in locations where demand is in closest proximity because of the cost to construct a water distribution system both to the recharge location and to the users of extracted water. Projected unmet municipal, industrial, and agricultural demands can be used to evaluate the proximity criterion.

Surface recharge facilities, recharge wells, and vadose zone wells tend to be located in areas with favorable hydrogeology, even though such areas may be far from where the water is ultimately needed. ASR wells tend to be located where the water is most needed during recovery, sometimes accepting less than ideal hydrogeologic conditions in the storage aquifer.

Landownership and Land Use. The landownership and general type of land use at and surrounding a potential MAR site should be identified during the planning process because it could affect the suitability of a given site. In general, a MAR facility is more likely to be developed if it is compatible with surrounding land uses. This is a more important criterion when considering recharge spreading basins versus recharge wells. Land use categories, including urban, agricultural, native and range lands, parks, critical habitat areas, and the location of inaccessible lands (such as military reserves), should be at a sufficient level of detail during the planning stage. Landownership and land use in the United States are available from the USGS (Homer et al. 2012).

In addition, information on whether a potential MAR site is located on public or private land could be a consideration when comparing potential sites, particularly when surface recharge basins are being contemplated for recharge. In the United States, public lands considered to be publicly accessible, and therefore potentially more amenable to MAR projects, include those lands managed by the following federal or state agencies:

- US Bureau of Land Management;
- US Bureau of Reclamation;
- US Forest Service;
- US Fish and Wildlife Service;
- State natural resources agencies, including the Division of Wildlife; and
- State land boards.

Cost. Cost should be assessed at the planning level to assist in comparing relative costs among sites, source waters, water treatment and conveyance issues, and recharge technologies. The cost evaluation should consider the anticipated capital costs for the items, including property; rights-of-way; and facility planning, permitting, design, and construction costs. It should also consider anticipated operations and maintenance costs, including water treatment, pumping, equipment repair and, ultimately, facility decommissioning costs. Land acquisition for spreading basins could be a significant capital expense because of the large areas needed compared with recharge wells. Construction costs for MAR projects using spreading basins may be lower than those for recharge wells. In confined aquifer settings, cost could be reduced substantially if existing wells could be retrofitted and used for recharge.

When there might be multiple potential sources of water, care must be taken to include an evaluation of the effect of its quality and the associated treatment costs, including the proper disposal of treatment by-products. At the planning stage, it may be sufficient to equate relative costs of water treatment at a given location to the concentration of several water quality constituents, such as total dissolved solids and nitrate. At more detailed stages of the project, there needs to be comprehensive analyses of common ions, metals, viruses, pathogens, and potentially of synthetic organic compounds to develop a more precise estimate of water treatment requirements and their costs. More detailed discussion of water quality issues is presented in Sections 5.2.5, 5.3.5, 6.5, and 7.4.

Site Access and Security. The potential MAR facility should be evaluated in terms of its ability to be accessed by system operators and to be secured against unauthorized entry. Site access may require crossing property not owned by the system operators, so formal access agreements should be secured at an early stage in the process. In addition, sites located in rural areas could be difficult to reach during inclement weather because of terrain or road quality.

Conditions Surrounding the Site. The area of the site should be of sufficient size to provide for facility construction, well drilling, maintenance equipment, storage of materials, and treatment facilities as necessary, including disinfection and filtration. Other conditions that should be considered include impact of noise on surrounding areas; conditions that could affect operations and use of nearby wells; presence of urban and recreational areas that could be affected by odors, insects, or noise originating at recharge sites; and areas located within the regulatory (100-yr) floodplain.

For confined and semiconfined aquifers, recharge wells and ASR wells typically fit within a small area; however, they store water within a radial distance of about 100–600 m (300–2,000 ft), depending on site-specific considerations. Land use within this radial distance likely needs to be controlled by the owner of the recharge facilities so that stored water is not pumped out by adjacent wells or contaminated by overlying land uses. A wellhead protection area or well field protection area (WPA) may be needed. Depending on state and local laws and regulations, this goal may be achieved by several means, including land acquisition, lease, and local institutional controls, such as ordinances that set aside this portion of the storage aquifer for protection of public water supplies.

4.2.6 Example Regulatory Criteria. This criterion describes the activities needed to obtain the necessary permits or fulfill other regulatory requirements for a MAR project. Regulatory and permitting issues that can be expected for a typical MAR project include acquiring water rights; obtaining permits for well recharge, including underground injection control (UIC) permitting; and obtaining permits relating to environmental impacts caused by construction of a facility and its mitigation. In addition, it may be necessary to obtain agreements from regulatory agencies that might have jurisdiction in the location where the MAR project or its impacts are anticipated. At the planning stage, this criterion should be used in a qualitative manner based on knowledge of the needed permits and regulatory requirements.

Regulatory requirements vary widely by state and as a function of site-specific features of an individual project (including the type of recharge, source water, and storage location); case studies can be particularly useful. See Appendix D for examples of case studies and seek out experts in the reader's region.

Water Quantity Laws and Regulations. The ownership of or the rights to use the water placed in underground storage, the right to store water in an aquifer, or the right to extract it are key regulatory issues a MAR project needs to address. In the United States, individual states have the legal and regulatory authority to govern water supply and use. This situation has created many variations on how water law is administered, ranging from permit-based systems to those based on a priority system that requires water court approval. A complicating matter is that most states regulate surface water and groundwater separately and often use different rules for each, so a thorough investigation of these requirements is needed. The natural resources or water resources agency within a state government within the United States generally administers rules for water supply and use. There may be separate regulations regarding water supplies for injecting and extracting water, and there may be a percentage of the recharged water that must stay in the aquifer.

In addition, in many states, there are additional regulations for certain aquifers that are at risk from existing or anticipated overuse. These aquifers include some of the groundwater basins in California, most of the Groundwater Management Areas or critical use areas in the coastal eastern states south of New York and in portions of Arizona, Designated Basins in Colorado, and others. Some states have specific rules regarding permitting for aquifer recharge or that discuss issues of underground water storage and capture. As of 2009, these states include Arizona, California, Colorado, Delaware, Florida, Idaho, Kansas, Nebraska, Nevada, New Jersey, New Mexico, Oregon, Texas, Utah, Washington, and Wisconsin (Hahn 2009, NRC 2008). Appendix C provides examples of regulatory requirements for MAR projects for selected states (Arizona, California, Colorado, Nebraska, New Mexico, and Texas). The laws and regulations listed for these states demonstrate the wide range of requirements.

Water Quality Laws and Regulations. Laws and regulations associated with water quality in the United States have mostly been developed at the federal level. The two key laws are the Safe Drinking Water Act and the Clean Water Act, both of which are administered by the US Environmental Protection Agency (USEPA). Many states have been granted primacy status to implement the federal rules and may have developed regulations that are more stringent. There may also be state rules or regulations regarding use of treated wastewater effluent as a source of water for recharge. Some states also have antidegradation policies for certain constituents, such as coliform bacteria and common ions such as sodium, iron, nitrate, and total dissolved solids. The environment or health departments within state governments generally administer rules for water quality.

The USEPA Safe Drinking Water Act (SDWA) has two provisions that apply to MAR projects. The SDWA has numerical water quality standards known as maximum contaminant levels (MCLs) that apply to water that is a current or future source for public supplies. Typically, MAR activities would not be able to diminish the water quality in an aquifer used for public supplies beyond the MCL numeric standards, including both the water used for recharge and any reactions that might occur within the aquifer caused by the addition of recharge water. If recharge is implemented using wells, then the SDWA's UIC regulations also apply and the recharge well or wells need to obtain Class V injection permits.

Another federal law that might apply to MAR projects is the Clean Water Act, also administered by the USEPA or states that have primacy for enforcement. This law states that there shall be no discharge of pollutants (regulated chemical constituents) into the nation's waterways beyond established maximum limits that would impair the beneficial uses of these waters. The Clean Water Act gave rise to the National Pollutant Discharge Elimination System (NPDES) regulation that is usually applied to industrial or municipal discharges. It would apply to MAR projects if the source of water being used for recharge is treated wastewater and also could be applicable for captured stormwater runoff.

In some states, such as California, state water quality legislation has had a large impact on how MAR projects are implemented. California's Porter–Cologne Act, referred to locally as the antidegradation policy, can restrict a MAR project from any water quality degradation above background levels, even if all constituents meet drinking water standards (Appendix C).

In 2013, the USEPA modified its policy with regard to UIC permitting of ASR wells, providing greater discretion to individual states for issuance of permits that reflect the needs, constraints, and opportunities for aquifer recharge in each state (USEPA 2013). Most importantly, the change in policy provides reasonable time and distance in the storage aquifer for natural treatment processes to occur that improve the recharge water quality, thereby meeting applicable water quality standards. Previously, UIC permitting provided no such provision. As discussed in Section 3.8, extensive research and experience have documented the natural ability of soil aquifer treatment (SAT) in unconfined and confined, unconsolidated and consolidated aquifers, to achieve improvement in water quality during aquifer recharge and storage.

Environmental Laws and Regulations. Environmental considerations are important and affect the amount of local support a project receives. Impacts to river flows and quality can affect adjacent riparian habitats. Land disturbed by construction of the MAR facilities, along with water distribution systems for source water and extracted water can negatively impact species and habitat. Normally recharge operations are perceived as benefiting the environment by making more efficient use of water when it is available without causing significant adverse effects. However, each project disturbs the landscape to some extent, which may have adverse effects on the ecosystem. Mitigation plans may be appropriate to address potential adverse environmental effects. The potential cost of implementing such plans should be incorporated in the project economic analysis.

In the United States, the environmental compliance process is based largely on meeting requirements of federal law. If a MAR project is supported with federal funds or uses federal lands, the project becomes subject to the National Environmental Policy Act of 1969 as amended (PL 91-190), typically referred to as NEPA. This act requires federal agencies to assess the environmental impact of a project, and for projects or actions expected to have a significant effect on the quality of the human environment, the proponent is required to prepare and file a formal environmental impact statement (EIS).

Some states and some other countries have enacted requirements similar to NEPA for projects supported by the state or requiring state permits (such as the California Environmental Quality Act, or CEQA, as described in more detail in Appendix C). In these states, projects are subject to the state environmental compliance process. For those projects supported by federal funds or requiring federal permits, both NEPA and state processes are in effect. Regardless of whether a formal assessment is required, the NEPA process is well developed, widely practiced, and can be used as a model for conducting environmental studies to evaluate environmental and social issues for MAR projects.

Addressing and evaluating the social, economic, and environmental issues and concerns of all the alternatives (including the no-action alternative) as early as possible should lead to a program that has wide support. To address the environmental and social issues adequately, it is necessary during the project planning phase to consider both the potential positive and negative environmental effects created by the project and mitigation required to overcome the negative effects. This consideration includes a detailed examination of the biophysical and socioeconomic consequences to adjacent properties and a thorough search for alternatives to avoid or minimize adverse effects. Procedures for accomplishing this type of analysis are essentially the same as those used in conducting environmental impact assessments and have been well documented in the literature (Jain et al. 1993, Canter 1996).

Even if the proposed project does not use federal funds or involve federal lands, an environmental impact assessment and a report can be important elements of a MAR project. Environmental impacts not only relate to the physical environment but also to the economic and social impacts of a project. The effects of the proposed project should include those that are (1) direct, (2) indirect, and (3) cumulative. Direct effects are related to a specific action, occurring at the same time and location. Indirect (or secondary) effects are also related to a specific action but are usually delayed or occur at some distance from the action. Cumulative impacts arise where an accumulation of small impacts from one event becomes large or where the addition of small but similar impacts from additional events create a large impact, such as expanding the number of surface infiltration basins. When the impacts appear to be minor and there is little or no opposition to a given project, then a preliminary environmental assessment included in the planning phase evaluation report may suffice.

Projects that disturb soils within channels and on floodplains are subject to US Army Corps of Engineers Section 404 permits. Other permits are required if wetlands will be adversely impacted.

Preliminary studies of environmental factors should collect all available data so that environmental problems can be discovered and addressed at the earliest possible time. Failure to do so can cause delays later in the project studies and result in a loss of public support. If rare and threatened or endangered species of animals or plants exist in the project area, early consideration should be given to possible mitigation measures so that all alternative plans can be considered along with their actual costs. If needed, the more comprehensive environmental impact studies should be conducted in conjunction with the engineering analysis during the design phase and the environmental impact report issued along with the design report.

An activity checklist can be useful in determining the types of environmental impacts that are associated with the proposed project. The activity checklist should include items that cover the entire lifecycle of the project and should include a narrative discussion of the environmental (physical, social, and economic) impacts of the proposed project. It should make a finding that the project will, may, or will not have a significant effect on the environment and describe mitigation measures, if required. Finally, the activity checklist should be accompanied by details of facility design, resources used, land acquisition, site access, construction, start-up, operation, monitoring, maintenance, rehabilitation, and ultimate disposition of the facility when the project is completed.

The activity checklist is also useful in developing a thorough understanding of exactly what is to be done, where and when it will be done, who will do it, as well as why it is being done. Furthermore, the alternatives associated with items on the activity checklist should be identified and included in the evaluation. The list should be updated periodically to incorporate changes in study directions and needs. The issues covered by an environmental assessment relate to water, ecology, land, social, and human elements, and to environmentally sensitive areas. Environmental and social issues vary with project location, design, size, and other variables. Items to be covered also vary depending on the governmental agencies involved in proposing or funding the proposed project.

Other Institutional and Regulatory Issues. Other institutional and regulatory constraints include a broad array of jurisdictional issues, customs, and activities that may affect a MAR project. These constraints include the following:

- Coordination among agencies, including the clear authority for a sponsoring entity to engage in MAR activities. For example, a surface storage entity may not have authority to do groundwater recharge; or requirements of local groundwater control districts may be inconsistent with, or overlap, those of state agencies;
- Lack of clear jurisdiction for any agency to perform groundwater recharge in a complex urban setting where many municipal and other agencies exist; and
- Issues regarding land use, including site easements and site access.

Solutions to institutional issues may involve establishment of a collaborative environment among all potentially affected entities. Creating partnerships among various groups and interests is an effective means of solving institutional problems. Example frameworks for coordination among governmental agencies are found in ASCE/EWRI Standards 33-09, 40-18, and 60-12.

Laws governing archeological artifacts and cultural resources must also be recognized. At sites where historic, prehistoric, or other cultural values may be found during excavation, special care needs to be exercised. When found, the proper government office should be notified immediately. Therefore, cultural resources investigations, including archeological studies, should be part of every project formulation, and appropriate mitigation measures should be added to the project, if they are found to be needed.

MAR projects using wells require smaller land areas compared with surface recharge projects, so the environmental requirement considerations discussed previously are often reduced for well recharge projects.

4.2.7 Data Gathering for Criteria Evaluation. Data collection and analysis are discussed in detail in Chapter 9, along with data organization and management. Some of the sources of available data for sites in the United States are presented in the previous sections associated with each criterion. Other sources of hydrogeologic information include state natural resources agencies and well drilling permits. Additional sources of water quality data include but are not limited to state and county health departments and annual reports of municipal water providers. It is recommended that available data be organized and evaluated in groupings that are consistent with the defined criteria to facilitate the site ranking process. A geographic information system (GIS) database is highly recommended for the storage and presentation of site-specific data.

A number of criteria have been presented that describe source water, hydrogeologic, environmental, and implementation considerations when evaluating the suitability of a potential MAR area or site. Many of these criteria could affect whether a MAR project is deemed feasible and could influence the selection of one candidate site over another. In many cases, the data for individual criteria are missing or incomplete. The hydrogeology and geochemistry of the target aquifer generally are the least characterized parameters, followed by the water quality of the source water. These basic uncertainties lead to a related uncertainty in the infiltration and aquifer recharge rates, residence time of the recharged water, geochemical reactions that might occur in the target aquifer, and water treatment costs. In addition, when data are available, they may not be of the needed quantity, location, or frequency to make accurate planning decisions. Historical water quality data that are too old may not be representative of the current conditions in areas that have experienced significant growth or changes in basin activities, such as increased pumping stresses and larger depths to groundwater.

With numerous criteria poorly understood, the greatest problem becomes making defensible and objective decisions on eliminating and retaining potential sites for further evaluation. This problem can be addressed in part by undertaking two activities: developing a sequenced or iterative approach to evaluating the potential sites and engaging stakeholders who may be potentially affected by MAR operations on the candidate sites. Stakeholders can be an important element of the public participation process by imparting site-specific knowledge and through their involvement and concurrence with the site selection process. Section 4.3.4 discusses public participation in greater detail. In general, phased development of MAR projects is recommended so that the lessons learned from earlier phases can be folded into the decision-making process for subsequent phases.

Case Study 12 in Appendix D is cited as an example of a project where public participation early in the design process, in the form of an outreach program, proved valuable in ultimate approval of the project.

4.3 DETAILED PLANNING: IDENTIFY FINAL POTENTIAL MAR PROJECT SITES

In many cases, the initial data analysis (Section 4.2) identifies multiple sites or areas where MAR projects may be feasible. An initial list of alternative projects then could be developed that combines potential sites and potential recharge methods. This list should be developed without regard to cost or other constraints that might otherwise eliminate potential projects, because many that are not considered to be viable may have initial interest and support from the public or other entities. A scoring and ranking process then should be undertaken to reduce the potential sites to a smaller number that might then undergo a more detailed evaluation. The scoring or ranking process should compare the project objectives with the site data as they have been organized according to selected criteria, as discussed in Section 4.2. This section discusses possible scoring and ranking methods for potential MAR projects.

4.3.1 Initial Project Site Screening. When a large number of potential MAR alternatives exist at the planning stage, it is recommended that they be assessed in a two-step process that consists of preliminary and detailed evaluations. The preliminary evaluation is intended to quickly screen out or eliminate certain locations and MAR technologies that clearly would not be feasible. It should be based on relatively simple and easy to apply criteria designed to quickly identify and eliminate potential projects that do not meet key project objectives or criteria. Even situations in which there are few candidate sites or recharge methods can benefit from conducting a two-step evaluation process to reduce the number that would undergo a more detailed evaluation. This step thus can be thought of as a fatal flaw analysis (OWRB 2010) or viability assessment (NWQMS 2009) for both potential MAR technologies and potential locations.

It is up to the project team to determine what criteria should be used in the preliminary evaluation and how those criteria should be used to retain or eliminate potential projects. The project objectives (Section 4.1) could be used as the basis for selecting the key criteria. In most cases, the hydrogeology and aquifer conditions in a candidate aquifer would be criteria included in a preliminary evaluation.

For example, Australia MAR guidelines recommend asking the following questions about a candidate site to assess whether it is viable. A negative response to any question would eliminate a candidate site from further consideration (NWQMS 2009):

- Is there sufficient demand for recovered water?
- Is there an adequate source of water available for allocation to recharge?
- Is there a suitable aquifer available for storage and recovery of the required volume?
- Is there sufficient space available for capture and treatment of the water?
- Does the capability exist to design, construct, and operate the MAR project?

In another example, in its statewide assessment of potential MAR sites, the Oklahoma Water Resources Board determined that a potential site had to meet all four of the following criteria or it would be eliminated (OWRB 2010):

- Seasonal variability in demand,
- Proximity of a site to areas of demand,
- Proximity of a site to water available for recharge, and
- Acceptable groundwater quality defined by TDS concentration.

The TDS criterion in this example is less important because many MAR projects store drinking water in brackish aquifers. The OWRB screening process continues with a threshold screening step if a large number of sites are retained after the initial step (OWRB 2010). The threshold screening consisted of assessing sites based on additional criteria and using a good/fair/poor assignment for each. Those sites assigned poor rankings for any criteria were eliminated, as were some sites that were assigned fair ratings. The criteria used for the initial screening analysis were used in the final detailed analysis of potential sites, along with criteria relating to source water availability and quality; aquifer storage, permeability, and residence time; and regulatory challenges (OWRB 2010).

4.3.2 Detailed Evaluation. Potential MAR projects that are retained from the preliminary screening evaluation should then undergo a more detailed evaluation. This step reduces the number of sites that are retained to a small number (or possibly to a single site) that would then be subject to additional field investigation and analyses prior to project design.

At this step in the evaluation process, a more quantitative approach is warranted. It is recommended that the evaluation criteria developed previously and used to organize available data be used for the detailed evaluation of retained MAR projects. A range of scores should be determined for all criteria, such as 1 to 3 or 1 to 10; the higher values mean that a particular project meets a criterion more favorably. The criteria should be defined by specific quantitative measures, and a range of scores should then be assigned based on those definitions. For example, if proximity of a project to demand were a selected criterion and the range of possible scores was 1 to 5, then proximity of less than 2 km (1.2 mi) might get a score of 5 and proximity greater than 10 km (6.2 mi) might get a score of 1.

In some cases, it may be necessary to define a particular criterion in different ways for different aquifer types or geographic regions being considered. For example, unconfined and confined aquifers may need separate definitions for criteria such as aquifer storage capacity, hydrogeologic suitability, residence time, and implementation costs because of the physical differences between these aquifer types. Such an approach was used in evaluating suitable regions for MAR projects over the eastern half of Colorado (CWCB 2007). Table 4-2 lists the criteria, their quantitative definitions, and how scores were assigned for each criterion for that study, including separate definitions for unconfined and confined aquifer settings for several criteria.

Deciding how to define each criterion quantitatively over a range of possible scores is likely to be an iterative process based on the available data, professional judgment, and feedback from the public and other stakeholders. The criteria scores should be defined as clearly and objectively as possible so that anyone reviewing the data and applying the criteria definitions would come up with similar scores. Although this process might require extra effort, the result is a scoring process that is viewed as transparent and unbiased and that leads to fewer concerns about the projects retained for further investigation.

The evaluation criteria selected for detailed evaluation of potential MAR projects are likely to cover a wide range of considerations and may not be of equal importance. If that is the case, then weighting factors should be assigned to each criterion, with larger weighting values representing the more important criteria. The weighted score for each criterion would be equal to its raw score multiplied by the weighting factor. As with the criteria scoring, defining weighting factors is likely to be an iterative process based on professional judgment and, possibly, stakeholder input. The goal of the weighting factors is to distinguish between relatively more and relatively less important criteria that are used in the detailed evaluation and scoring of potential MAR projects. It is helpful to review the project objectives when defining weighting factors to make sure that one is consistent with the other.

It may be beneficial to define criteria scores and weighting factors so that the maximum possible score for a potential project is 100 or some other commonly understood value. This definition may facilitate understanding of the scoring and ranking results by the project owner, the stakeholders, and the public.

Once criteria scores and weighting factors have been assigned, then the next step in the detailed evaluation phase is to assign a score for each criterion for each potential project, multiply that score by the criterion's weighting factor, and then sum all the weighted criterion scores to obtain a total score for each project. The projects can then be sorted by score; those receiving a higher score are more favorably aligned with the evaluation criteria and thus more likely to be successful projects. The project scores can be further categorized by geographic region, aquifer type, recharge method, or other factors to assist in evaluating, presenting, and explaining the results.

In the Australian MAR guidelines (NWQMS 2009), the second evaluation step involves categorizing the degree of difficulty against 14 criteria to gauge the extent of investigations required to complete risk assessments for regulatory approval at any prospective MAR site. This categorization can be used to rank alternative MAR projects at an early stage.

Table 4-2. Example Scoring Measures.

Evaluation Criteria	Criteria Description	Scoring Measures: High (10 9 8)	Scoring Measures: Medium (7 6 5 4)	Scoring Measures: Low (3 2 1)
Hydrogeologic considerations				
1. Aquifer storage capacity	Available capacity for recharge	>0.6 m^3/m^2 (2 AF/ac)	0.1–0.6 m^3/m^2 (0.25–2 AF/ac)	<0.1 m^3/m^2 (9 < 0.25 AF/ac)
2. Hydrogeologic suitability • Unconfined aquifers • Confined aquifers	Potential rate of aquifer recharge —Estimated from aquifer K values —Estimated from aquifer T values	>10 m/day (>250 ft/day) >300 m^2/day (>900 ft^2/day)	1.0–10.0 m/day (50–250 ft/day) 100–300 m^2/day (300–900 ft^2/day)	<1.0 ft/day (<50 ft/day) <100 m^2/day (9 < 300 ft^2/day)
3. Residence time • Unconfined aquifers • Confined aquifers	Duration recharged water is in aquifer Subcrop proximity to alluvial aquifers	>1 year >5 km (>3 mi)	4 months–1 year 1.6–5 km (1–3 mi)	<4 months <1.6 km (<1 mi)
Environmental considerations				
4. Water quality	Aquifer water quality with respect to state standards, soil leaching potential	No standards exceeded; minimal leaching potential	Limited areas where standards exceeded; minor leaching potential	Large areas where standards exceeded; strong leaching potential
5. Habitat concerns	Presence of threatened and endangered species habitat; effect on wetlands	Minor area of T&E habitat; no effect on wetlands	Some T&E habitat; some wetlands affected	Much T&E habitat; wetlands affected
6. Waterlogging and nonbeneficial use	Potential to create high water table and increased ET by phreatophytes	Low concerns for waterlogging effects	Medium concerns for waterlogging effects	High concerns for waterlogging effects
Implementation considerations				
7. Landownership and land use considerations	Proportion of area with accessible public land, multiple jurisdictions	Many areas of public and nonurban land	Some areas of public and nonurban land	Mostly private and/or urban land
8. Existing infrastructure	Proximity of infrastructure (pipelines, ditches, etc.) and available capacity	Suitable infrastructure <10 km (<6 mi) from area	Suitable infrastructure 10–30 km (6–20 mi) from area	Suitable infrastructure >30 km (>20 mi) from area
9. Proximity to areas with demand	Recharge areas nearby to areas of projected unmet demand in 2030	Near areas with demands >4,000 m^3/year (>10,00 AF/yr)	Near areas with demands of 2,000–4,000 m^3/year (5000–10,000 AF/yr)	Near areas with demands <2,000 m^3/year (<5,000 AF/yr)
10. Implementation costs • Unconfined aquifers • Confined aquifers	Relative land costs for construction Depth to aquifer and proximity to existing high-capacity wells	Low cost <75 m 9 (<250 ft); many wells in area	Medium cost 75–300 m (250–1,000 ft); few wells in area	High cost >300 m (>1,000 ft); no wells in area

Note: Criteria 2, 3, and 10 have separate definitions and scoring measures for unconfined and confined aquifers.
Source: Adapted from CWCB (2007).

It is recommended that scores be reviewed carefully to identify whether there may be potential biases in the criteria definitions used in scoring or with the weighting factors. A closer review of the available data, changes to criteria definitions, and adjustment of weighting factors may all be needed to obtain an objective evaluation process.

4.3.3 Selection of Sites for Feasibility Studies. A list of potential sites that have been selected for more detailed analyses can be developed after careful evaluation and scoring as described previously. It is likely that sites on this list will require further detailed feasibility studies, including field studies and pilot testing. The purpose of the feasibility studies will be to address site-specific questions that might exist about site hydrogeology, environmental, and implementation considerations. Once the feasibility studies are completed, it may be appropriate to reassess the potential MAR sites using the scoring criteria. This can be an iterative process until the group arrives at the most favorable site and technology.

4.3.4 Public Participation. MAR projects often result in impacts to various groups of people both in the immediate area and at a distance from the recharge site. These impacts can range from a change in the land use and level of activity at the site, to a change in the groundwater level, to a change in the water quality of groundwater in the receiving aquifer, to a change in the flow at adjacent streams and springs, and to other environmental impacts.

It can be very beneficial to involve the public and other potential stakeholders during the planning process to inform them of the proposed project, describe how managed aquifer recharge works, tell them how they will benefit, and explain the safeguards used in the process. Forums should be established to receive suggestions and comment, to identify their concerns, and to work with the forums to address the concerns. These interactions can help avoid misunderstandings or lack of support for the selected projects that could otherwise take large efforts to overcome at a later date.

An advisory group can provide a communication link between the project proponents and the public. Members of an advisory group should represent the many elements of the public in the affected areas and might consist of members of the public and representatives from business and environmental groups, service organizations, and governmental agencies. The advisory group should meet frequently enough that feedback on project progress and issues can be received on a timely basis. A separate public education program should be considered as a way of preventing public misconceptions. Public input and discussion can result in realistic modification, rejection, or acceptance of the proposed projects.

4.3.5 Documentation. The project evaluation process should be presented in a report, or set of reports, that documents the purpose of the analysis, the project objectives, the study methods, data sources, and results. Depending on the number of potential sites that have been part of the planning process, it may be appropriate to develop separate reports for the initial site screening and for conceptual design.

Initial Site Screening Report. This report summarizes the initial planning activities and findings, including a clear description of the reasons to support the project or projects that have been selected for additional site-specific feasibility study testing and analyses. At a minimum, the site evaluation report should include the following information:

- Objectives and criteria used in evaluating potential projects;
- Summary of the data used, results of its analysis, and appropriate references;
- Input obtained from the public advisory group;
- Projects selected for additional study;
- Advantages and disadvantages of potential projects;
- Recommendations for additional data and field studies that should be made; and
- Possible regulatory issues that need to be addressed.

Depending on the study, there also may be a need to document why potential projects were rejected. In addition, it may be advantageous to include a discussion of how the retained projects would fulfill the project objectives compared with a future scenario without a MAR project; the latter is referred to as the "no-action" alternative.

Conceptual Plan Report. The information on the recharge processes being considered and the characteristics of the one or more alternative sites that have been identified should be organized showing the advantages and disadvantages of each site. From this base, combinations of recharge methods and sites can be developed into conceptual plans to be evaluated for their ability to achieve the recharge objectives. The conceptual plan should be a generalized layout and plan of the ultimate recharge and recovery facilities with additional detail on locations for test facilities that could be constructed to confirm the conceptual plan. The conceptual plan should include the location and sizing of basins, levees, recharge and recovery wells, pipes, pumps, buildings, controls, and other facilities in sufficient detail that an order-of-magnitude cost estimate can be developed for comparative purposes. The cost estimate should provide an estimate of unit costs for recharged and recovered water, or unit costs for expansion of peak water supply capacity, if this is one of the recharge objectives.

A key element of the conceptual plan is whether to use existing land or facilities that may be available or to construct new facilities. Quarries, gravel pits, abandoned wells, and other such facilities may be available and should be considered for test program purposes. Such existing facilities are not always situated appropriately and are not always in a condition suitable for recharge purposes. Therefore, it is important that program objectives not be compromised by a decision to use existing facilities that are unsuitable for the purpose. Once a recharge facility becomes severely plugged and the program loses support, it is difficult to regain support by pointing out that further investment in test facilities would more likely achieve success. These problems can severely hamper the conduct of a successful recharge program. Where existing basins or wells are used for testing, great care has to be exercised to identify and, if possible, correct shortcomings in design or construction of the existing facilities that could adversely affect the recharge program. Research and further investigations play an important role in avoiding such problems.

CHAPTER 5
DESIGN AND TESTING

Previous chapters of this standard discuss MAR projects from a broad conceptual level, including general hydrogeologic concepts and guidance on how to evaluate potential MAR sites and technologies. This chapter focuses on items to consider in the design of surface (Section 5.2) and subsurface MAR projects (Section 5.3), pilot project design (Section 5.4), and design documentation (Section 5.5). Where appropriate, specific design criteria are suggested. It is assumed that site selection and feasibility-level field testing (Section 4.3.3) have been completed and that the proposed project has been approved. Construction, operations and maintenance, and closure of a MAR site are presented in Chapters 6, 7 and 8, respectively. Data collection necessary for all phases is discussed in Chapter 9.

Because of the complex nature of MAR projects, the design process is likely to evolve concurrently and iteratively from the conceptual level through the more detailed phases of final design and eventually construction (Figure 5-1). Components of a MAR project include delivery systems, recharge systems, monitoring systems, and ongoing operations and maintenance (O&M) plans. A robust conceptual design should be flexible enough to allow for what is discovered at the project site during the feasibility analysis and early design phases. Final design of some components should depend on actual construction of others. Because of the specialized and iterative nature of MAR facility design, it is advisable to engage experienced MAR design consultants and construction contractors to guide the process.

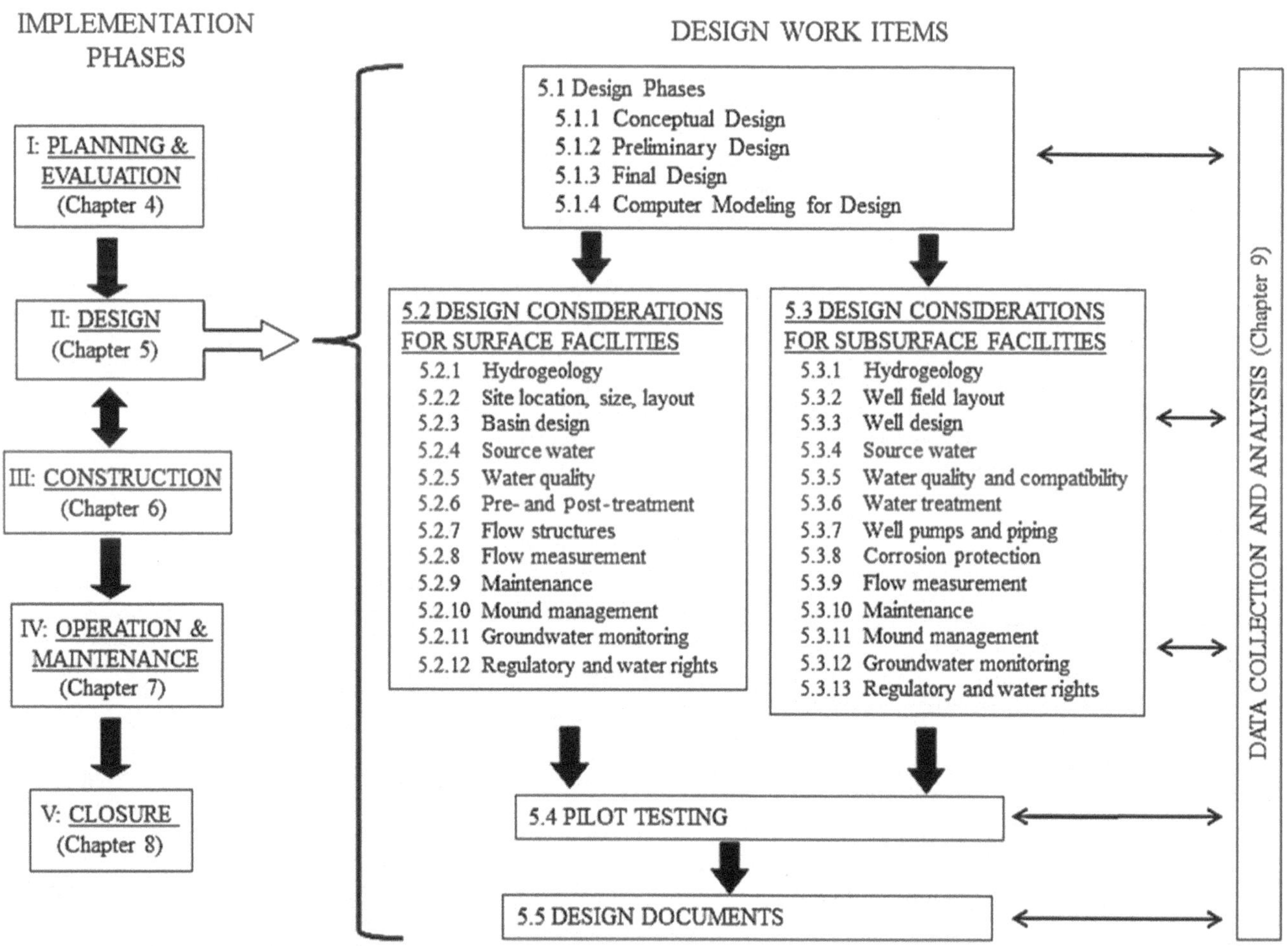

Figure 5-1. Details of the design phase of a MAR project.

MAR projects can comprise a wide range of infrastructure and design components, including but not limited to wells, basins, levees, canals, water diversion structures, water treatment facilities, pipelines, electrical power lines, metering and control facilities, monitoring, supervisory control and data acquisition (SCADA), and telemetry systems. The detailed design of these facilities typically requires a team of civil, mechanical, geotechnical, electrical, instrumentation and control (I&C), process engineers, and hydrogeologists. Operators and those tasked with performing maintenance operations should be involved with the design process because this involvement is likely to result in process improvements. This team should be managed in a coordinated fashion so that each team member clearly understands the objectives of the project (Section 4.1.1), the project's performance criteria (Section 4.2), and how the project is intended to be operated and maintained. For example, some MAR projects are only operated seasonally or intermittently, which introduces operational challenges and risks that should be considered and explicitly addressed during the design process. Facilities used intermittently should be designed with capability to be drained and prepared for winter, to remain inactive for extended periods of time, and/or to start up or shut down within short periods of time. As such, it is critical that the preliminary design process adequately identifies the design criteria and characterizes the types of considerations that should be addressed in the final design phase.

5.1 DESIGN PHASES

One of the primary design objectives of any MAR facility is to minimize the cost per unit volume of recharge, storage, and recovery over time (such as in dollars per cubic meter). For an aquifer storage and recovery (ASR) well or well field, an additional primary objective is to minimize the cost per unit volume of water recovered per unit of time (such as dollars per cubic meter per day). Both criteria are equally important for ASR wells, most of which are seasonal operations that are often compared with desalination treatment plants, and long pipelines that may be operated year-round, producing larger volumes of water but at much higher capital cost.

An integrated and phased design plan is critical for project success. Plans and specifications should be prepared in phases, allowing for a progression in detail, combined with review by stakeholders and regulatory entities, as needed. The process can commence during the planning stage and proceed to construction once uncertainty is reduced to an acceptable level. For instance, the final design of downhole pumps and valving for ASR wells, system controls, and wellhead plumbing, and metering should not occur until after the wells have been constructed and aquifer testing has been completed. Design documents for the various stages are discussed in more detail in Section 5.5.

5.1.1 Conceptual Design. During planning (Chapter 4), conceptual designs consist of descriptions of facilities with general components, dimensions, volumes and/or rates, and operating scenarios. They should also contain data collection elements, such as monitoring wells and flow monitoring structures. The intent is to describe how a MAR facility might be developed given known objectives, hydrogeologic conditions, source water options, and water recovery operations. During this design phase, it should be appropriate to develop a flexible approach that can be modified to adapt to physical conditions and regulatory requirements at the proposed MAR site.

5.1.2 Preliminary Design. Preliminary plans and specifications evolve from conceptual plans (Chapter 4). Preliminary design should occur after the detailed planning phase (Section 4.3) and any needed feasibility studies have taken place. The selected plan is developed in sufficient detail to include site diagrams, schematics, and layouts of the early project configuration that are feasible from geological, hydrological, engineering, environmental, and economic points of view and that address applicable laws, regulations, and water rights. Content should be at a level that allows cost estimates to be prepared and used in cost-benefit analyses. Pilot testing should be appropriate after this design phase to provide needed input to cost-benefit analyses of certain design elements (Section 5.4).

5.1.3 Final Design. After the planning and preliminary design phases are completed, the MAR project transitions to detailed design. Here, detailed design plans and specifications for construction of the infrastructure and improvements associated with the full-scale project are developed. Although other individuals and disciplines (including engineering and construction professionals) take the lead, it is important that the individuals involved in earlier planning stay involved for design continuity.

Before initiating detailed design of a MAR project, the owner should decide how the project, or projects, will be delivered. In other words, will the project be delivered using a conventional design-bid-build approach, or will an alternative delivery approach, such as design-build or construction manager-at-risk (CMAR) approach be utilized? Selection of a preferred project delivery method is a function of the scale of the project, the schedule objectives, the owner's desired level of involvement and decision making, and the owner's willingness to accept and manage risk. The choice of project delivery method drives how design is executed and affects the level of detail required for construction plans and specifications.

The final design is the last stage that produces documents to be used to construct a MAR project. It should be done with the detail necessary to permit an accurate estimate of costs, considering the ease of construction and operational procedures. Final plans and specifications are prepared from the draft documents after further data collection, analysis, and input from stakeholders and regulators. The documents need to be prepared with sufficient completeness and detail for construction activities. The documents include sufficient construction drawings and specifications so that the materials and equipment needed to construct the MAR project can be identified.

5.1.4 Computer Modeling for Design. Because of the complexity of many design criteria, computer modeling is a useful design tool. Computer modeling can be done throughout the planning and design phase to evaluate design options, operations, and management scenarios. A model is a sophisticated method of data analysis that simulates the interactions of related variables. Modeling can reveal sensitive variables that influence project feasibility and drive data collection efforts. Simple numerical groundwater models consider important aquifer boundary conditions and variable water delivery schedules. Sophisticated numerical models can include multiple aquifer layers; aquifer heterogeneity; complex boundary conditions; anisotropy or other variations in hydrogeologic parameters; the influence of surface water features such as lakes or stream; and alternative delivery options such as average year, dry year, or wet year flows.

Models can be used to guide surface design by estimating mound height and area beneath ponds and potential effects on nearby features. Estimates of MAR well recharge and withdrawal rates are critical design criteria. Modeling can be used to estimate head buildup rates and assist in estimating potential well-to-well interference and cumulative impacts of multiple MAR projects. The magnitude and direction of interference is a function of the distance between wells, aquifer properties, and the rates of both recharge and withdrawal. Geochemical modeling, based on mineralogy of aquifer materials and quality of recharge water, is useful to reveal the potential for adverse changes in water quality or clogging and hence to identify necessary pretreatment. Geochemical modeling is best conducted based on analysis of cores, not drill cuttings. Use of models in MAR project design and evaluation is discussed further in Section 9.6.

5.2 DESIGN CONSIDERATIONS FOR SURFACE FACILITIES

General design considerations for surface facilities fall into the following categories and are discussed separately:

- Hydrogeology;
- Site location, size, layout, and operational flexibility;
- Basin design;
- Source water;
- Water quality;
- Pre- and post-treatment in sedimentation basins;
- Flow structures;
- Flow measurement;
- Maintenance;
- Mound management;
- Groundwater monitoring; and
- Regulatory and water rights.

5.2.1 Hydrogeology. The hydrogeology of the site is the most important technical criterion when siting a MAR surface recharge facility. An extensive geotechnical investigation should be conducted at the site before preparing detailed plans and specifications. Factors to consider in the geotechnical investigation are discussed in Sections 4.2.3 and 9.4. Because the subsurface can be highly variable, numerous borings should be advanced within the proposed footprint of each basin to adequately characterize the depth, stratigraphy, and character of unsaturated materials beneath the facilities and the underlying aquifer. ASTM Standard D420 provides more information on site characteristics needed for engineering design and construction purposes.

Accurately estimating hydraulic parameters, such as basin infiltration rates, vertical permeability, aquifer transmissivity, and porosity, is critical to project success because they are the primary drivers that determine the wetted area necessary to achieve the project objectives. Hydraulic parameters also affect the formation of recharge mounds (discussed later), which, if they rise too close to the ground surface, can affect recharge operations until the mound has time to dissipate.

Aquifer testing is used to measure hydraulic flow and storage parameters in a target recharge aquifer and to assess infiltration rates. The standard practice described in ASTM D5270 and D5786 requires constant-rate pumping to produce measurable drawdown in both the pumping well and, ideally, in one or more nearby observation wells. Drawdown measurements collected during pumping are then analyzed to determine aquifer transmissivity, leakance, and storage coefficient values.

Throughout and especially at the end of the constant-rate test, water quality samples should be collected and analyzed to establish the ambient water chemistry and to assess whether key parameters change during pumping.

5.2.2 Site Location, Size, Layout, and Operational Flexibility. Because land and infrastructure costs can be high, MAR surface recharge site location, size, and layout are critical to project success. Most importantly, the site should have favorable hydrogeology and good access to source water. Surface facility design needs to reflect anticipated recharge volumes and rates and source water availability. Design depends on whether the facility is in-channel or off-channel and should take into consideration native flows and hydrologic variability and flood potential. The shape of the ponds and the water depths that will result in maximum infiltration rates will depend on the water availability to the site and the spacing of ponds. Site design considerations may also include identifying areas that should not be disturbed during construction, such as areas with wetlands, sensitive vegetation or habitat, and areas where access should be restricted.

Surface MAR facilities are best located on large, flat sites with permeable soils and a sufficient unsaturated zone to allow water to infiltrate vertically without causing excessive mounding. Site design may include adding soil amendments. A soil amendment is any material added to a soil to improve its properties, such as water retention, permeability, water infiltration, drainage, aeration, and structure. The most common use of soil amendments is to provide better growing conditions for plants. Adding soil amendments to improve infiltration rates is generally cost-prohibitive and risky. There may be some utility in adding soil amendments, such as gypsum or low pH water, to address changes that may have occurred in the target strata because of precipitation or cation exchange between the recharge water and target strata sediments. Extensive site-specific research would be required to assess whether this approach would work and be cost-effective.

Highly permeable vadose zone sediments are optimal to allow recharge water to move away from the spreading basins. Recharge volume and rates are proportional to the wetted area and therefore site size. However, grading is costly, so relatively flat sites with homogeneous, permeable soils, such as river bottoms or where any fine-grained surface soils have been removed, are preferred. If the objective is to recharge 1 million m^3 per year [810 ac-ft/year (afy)], and the basin wet/dry operations ratio is 80% with an average recharge rate of 0.2 m per day (8 in./day), then approximately 17,120 m^3 [1 million m^3/(365 days × 0.8 × 0.2 m/day)] or approximately 4 acres) of basins are required. Additional area is necessary to accommodate a pretreatment sedimentation pond and administrative infrastructure. The amount of buffer lands around a MAR facility should also be taken into consideration to address local concerns about noise and odors, particularly if reclaimed wastewater is being recharged. The site layout should also provide operational flexibility and critical maintenance, including access and storage of heavy equipment.

Ultimately, the size of the site depends on how efficient the facility is at maximizing the wetted area of the facility. Basin clogging is inevitable, so the facility should be designed to allow the basin to be periodically dried and the clogging layer removed, or to have multiple basins within the facility so that one or more basins can be dried and cleaned while others are in use. A realistic wet/dry ratio determined in a pilot study (Section 5.4) should be a primary design consideration.

The use of multiple smaller basins instead of fewer larger ones allows the following:

- Wet/dry cycling to occur year-round, which is operationally cost-effective;

- Optimizing deliveries based on basin-specific stage versus infiltration rate relationships; and
- Optimizing basin-specific treatment (including for algae and mosquito control).

Traditional terraced agriculture is certainly one of the most common surface recharge methods in arid areas, particularly in the Middle East. Where the terraces are well maintained, they effectively control runoff and improve aquifer recharge but, once allowed to fall into disuse, they progressively lead to gully erosion, collapse of the retaining walls, destruction of the whole system, and severe modification of the hydrological regime. Therefore, whatever the economic virtues of such terraces, it should be recognized that their abandonment on a large scale can upset the hydrological conditions within a basin for a considerable time (NRMMC et al. 2009).

5.2.3 Basin Design. The objective of basin-type recharge projects is to obtain the maximum ratio of wetted area to gross land area. Typical ratios for existing basins are about 75%; however, in urban areas the ratio may approach 90%. Conditions of service and maintenance also place some limitations on their use. In general, the coarser the texture of the surface deposits, the higher the initial and sustained infiltration capacities are. Target wetted area must be designed to accommodate the target recharge rate given local geologic conditions and availability of source water.

The results of the geotechnical investigation, along with an understanding of the availability and timing of source water supplies, guide the development of customized designs for each facility, including depth of excavation and basin configuration, and define the extent of excavation and thickness of permeable fill material, if required. In areas where flows are intermittent, it may be important to maximize capture of source water, which would lead to having deeper basins, even if this option reduces overall infiltration rates. Where source water is available on a more consistent basis, it may be important to maximize infiltration rates, which would lead to design of shallower basins that can be cleaned easily and frequently. Basin bottom elevation must be designed to consider water mounding during recharge, as determined by evaluation of sediment and aquifer characteristics and modeling (Section 9.6.2). Basin bottom elevation design should also take into consideration geologic conditions, such as local confining layers above the target aquifer. Overexcavation may be necessary to remove native layers with low permeability and followed, if needed, by backfilling with permeable material to provide sufficient height above the water table to allow mounding to occur.

In undeveloped areas, levees or berms often are constructed to retain surface water flows that are used for recharge. In many cases, they can be constructed by bulldozing existing sediments into place, without regard for fill slopes or compaction. However, in an urban area where seepage or levee failure may damage private property, greater attention should be given to foundation and embankment construction. In general, levees can be constructed with side slopes of 1.5 to 1, with an allowance for freeboard ranging from 0.3 to 0.9 m (1 to 3 ft), depending on the compaction of the material, size of basin, and wind direction. Selection of levee material depends on anticipated longevity of the facility, with concrete having the maximum durability. Roads are usually constructed on the levees for patrolling, inspection, operation, and maintenance. Section 3.5.1 discusses dikes, inflatable rubber or fabric dams, and dikes.

5.2.4 Source Water. The source water used for MAR facilities includes raw or treated water delivered from various sources at various rates and times. These waters may include treated or untreated lake or river water from the same watershed, imported water from another watershed, treated potable water from either surface or groundwater sources, treated urban stormwater, or reclaimed wastewater from sewage treatment plants or industrial operations such as coal seam gas production. These sources may be available separately or used in combination. Facility design needs to carefully balance the delivery, recharge, and storage volumes at appropriate scales. This balancing is done by conducting a source water availability assessment (Sections 4.2.2 and 9.5.1) and mass balance model (Section 9.6.1) to facilitate design of conveyance and storage features. Storage options include recharge basins, settling basins, or off-site reservoirs.

In-channel facilities used to recharge infrequent precipitation events of short duration would be designed differently than a facility that receives water from a pipe where the owner can use storage at the supply end and simply open a valve when available. A source water delivery analysis (Section 9.5.1) on a daily scale is necessary to size facility diversion structures, headgates, delivery canals, ponds, and outlet works properly.

The delivery of reclaimed water sometimes uses other designed methods of delivery that places environmental buffers between the source and the facility. For instance, instead of being pumped directly from a wastewater treatment plant, the delivery of reclaimed water may be more accepted by the public if delivered through a river, canal, or wetland. Each approach requires different design considerations.

5.2.5 Water Quality. Design of surface facilities needs to take into consideration water quality conditions of the source water to minimize clogging. The total suspended solids or turbidity of the source water should be characterized and incorporated into the design of the surface recharge facilities. Water chemistry also needs to be evaluated for possible reactions that might lead to rapid clogging or to adverse geochemical reactions once the infiltrating water reaches the underlying aquifer. Other site-specific conditions should be assessed that might lead to sources of clogging agents. An example is potential windblown sediments from nearby sources, including basin walls. Monitoring of changes in water quality may be a necessary element of the site design.

Health effects from pathogens associated with irrigation and nonpotable municipal wastewater used for MAR is of concern. The greater the human exposure, the lower the allowable concentration of pathogens should be. Where human exposure is significant, levels of indicator organisms, including coliform bacteria, viruses, and parasites, should be essentially zero (USEPA 1992). For crop irrigation, however, there is also concern that undesirable organic chemicals can be absorbed by plants and ingested by consumers. Total organic carbon contents in the water after MAR usually are on the order of a few milligrams per liter (mg/L), and concentrations of individual compounds may be on the microgram per liter (μg/L) level (Bouwer et al. 1984). These concentrations are very low, but some organic compounds may accumulate in the soil because of adsorption. The questions then are, will those chemicals be transmitted by the groundwater and absorbed by crops, and how will they affect humans or animals consuming those crops? There are risks associated with pathogenic organisms in surface waters when applied for irrigation and in groundwater recharge projects (Lee and Jones-Lee 1995b, 1996; Nellor et al. 1984).

5.2.6 Pre- and Post-Treatment in Sedimentation Basins. All water used for recharge should be as free of suspended solids as possible to minimize basin clogging. Depending on the source water quality, some pretreatment may be necessary. If the source water contains high levels of sediment, as might occur from a surface water source during flood flows, part of the flow may need to be bypassed, typically until flows decrease from their maximum, or sedimentation basins may be needed as pretreatment before the water is routed to the recharge basins. Projects with water sources in which turbidity and total suspended solids (TSS) concentrations vary over a large range may need to have water quality thresholds established so as to prevent the diversion of water that could cause excessive clogging. Developing these thresholds requires a good understanding of how TSS loading affects facility performance, the cost of cleaning the facilities, and the value of the water lost or gained.

Sedimentation basins should be designed to allow adequate time for sediment to settle and should be small enough to minimize evaporative losses. Unlike the recharge basins, pretreatment basins need only infrequent drying and removal of accumulated sediment, but their design must allow for this activity. Extended periods of time are required to settle suspended particles without the aid of flocculants or coagulants. The addition of flocculants or coagulants, which include a wide range of chemicals such as alum, ferric chloride, lime, and organic polymers, has the potential to enhance the settling rate and reduce the basin volume necessary for adequate detention. A United Nations Report (UN 1975) notes that water detained for 2 to 3 days at a constant 5 m (16.4 ft) water depth reduced the suspended solid load by 300 to 1,000 mg/L, whereas the proper dose of chemical coagulant achieved the same results in a few hours. Miller (1980) found that either alum or iron sulfate at 2 to 40 mg/L was effective for "turbid" surface waters and that 40–300 mg/L was required for effluent from wastewater treatment operations using sedimentation and activated sludge treatments.

More recent work conducted by the Orange County Water District (OCWD) in Orange County, California, found that whereas chemical pretreatment was successful in reducing total suspended solids concentrations in the source water, the presence of residual chemicals in the product water caused rapid clogging in large test basins filled with native sediment. It is thought that the residual chemicals reacted with the naturally occurring fine-grained sediments in the test cells, causing clogging rates to be elevated. This finding highlights the need to conduct site-specific studies before design decisions are made.

Post-recovery water treatment can include disinfection, filtering, pH adjustment, membrane treatment, and/or softening if the recovered water is anticipated to be hard. If recovered water is to be used as a source of supply to a potable water system, then local agencies that have jurisdiction over new source approvals and regulate drinking water systems should be involved in establishing post-recovery water quality criteria and determining treatment requirements.

The selection, design, and construction of water treatment facilities require specialized expertise and thus should be performed by licensed professionals with extensive experience in water treatment. It is prudent to consult with regulatory agencies early in the MAR project planning phase to identify data that must be collected during the construction process to evaluate the need for, and type(s) of, water treatment.

5.2.7 Flow Structures. If the recharge water is obtained from a river or lake, headgates and associated canals need to be part of the design to convey the source water to the recharge facility. Gated culverts of adequate size located through levees have been used successfully, as have weirs and spillways. Design consideration should be given to preventing erosive velocities on the downstream side of such structures. If water enters via pipeline, those water lines need to be sized properly to accommodate future growth. Overflow structures should be designed to handle routine drainage and flooding. For multiple-basin systems, adequate control of flow into and between the basins should be provided. Each basin should be hydraulically independent to allow optimal cleaning and maintenance of individual ponds and thereby improve MAR facility efficiency.

5.2.8 Flow Measurement. Flow measuring devices are necessary at the inflow and outflow location for each recharge facility for recording and regulating flow rates. Flow measurement may also be necessary between basins to evaluate individual basin performance and to design routine maintenance practices, such as wet/dry cycling. Both instantaneous and cumulative flows should be measured and recorded regularly. Reliable flow-measuring devices come in many models and types, so the choice is dependent upon the specific MAR project. Each recharge basin should also be equipped to measure water depth (stage) on a continuous basis. Basin volumes should be calibrated to dedicated stage-measuring devices. Because stage change over time is a key operating parameter, designing redundant measuring devices allows critical parameter data validation and verification. As with any project, the use of multiple methods to collect basic data is helpful in that it provides redundant and independent datasets, which allow for data verification and provide a backup system in the event of failure of one system. For example, a recharge basin could be equipped with a staff gauge that is read manually and a submersible pressure transducer or orifice line.

Evaporation rates are an important component of the mass balance equation for a surface recharge facility. Evaporation rates may be estimated either by measurements from site-specific monitoring equipment or through use of nearby meteorological data. Numerous methods and equations exist to estimate evaporation rates, including Penman, Penman–Monteith, and Priestley–Taylor, described in many hydrology textbooks.

5.2.9 Maintenance. An O&M program should be designed for any MAR facility, as discussed in Section 5.5.2 and in greater detail in Section 7.2. O&M should focus on maintaining recharge capacities and facility efficiency at the lowest cost.

Clogging is inevitable for any surface recharge facility. The design of surface facilities should allow for routine maintenance to limit and control clogging caused by suspended solids or algae and bacterial growth. Design considerations include adding additional basins to allow chemical treatment (spraying or biocide) and wet/dry cycling and to allow periodic maintenance. Bacterial growth can be sensitive to pond stage height and residence time. At an increased stage, anaerobic bacterial growth can accelerate and significantly reduce infiltration rates and wet/dry cycle times.

5.2.10 Mound Management. The design of a surface MAR facility must take into consideration the anticipated recharge rate and the amount of available hydraulic head to work with. It should also take into consideration seasonal or long-term changes in water levels in the aquifer in the general vicinity of the proposed project. When recharge rates exceed aquifer capacity, there is a potential for waterlogging soils or even flooding the surface facilities or nearby structures. Mound management scenarios can be evaluated using analytical or computer models (Section 9.6). Design measures to mitigate these effects include determining groundwater levels that will not be exceeded, barrier trenches, additional site drainage, and conveyance systems to relocate ponded water to areas not experiencing high groundwater.

Higher hydraulic heads associated with greater water depths in a surface recharge system can produce higher infiltration rates, but they can also tend to compress clogging layers once they form. Greater water depths provide more bank area for infiltration, increasing infiltration rates. Greater water depths also can reduce the rate of turnover of the water, thereby tending to affect water temperature and increase algae growth rates. Shallow basins have a higher rate of turnover of the water in the basin, which reduces the growth of suspended algae. But shallow basins have higher water temperatures in hot climates, thereby promoting algae and aquatic plant growth rates and also increasing evaporative losses.

5.2.11 Groundwater Monitoring. From a design perspective, monitoring is necessary to evaluate the aquifer's response to recharge compared with the design targets and to evaluate project operations. If possible, existing monitoring wells should be used; otherwise, installation of shallow monitoring wells should be considered, using methods recommended in ASTM Standards D5092 and D5521. Nested monitoring wells should be located near recharge basins to evaluate vertical gradients and the aquifer response and in areas where an excessive water level rise could cause flooding or waterlogged soils and possible liquefaction. Monitoring wells should be placed on the facility perimeter to evaluate hydraulic gradients and water quality. These data can be used to evaluate and improve facility performance, possibly through the use of computer models (see Section 9.6.3).

Monitoring wells are installed using the same drilling and construction techniques as those used for recharge wells. However, monitoring wells are typically smaller in diameter and are completed with polyvinyl chloride (PVC) casing. Monitoring wells should be drilled to the same depth and screened over the same zone(s) as recharge wells to enable collection of representative water level and water quality data. Depending on the depth and diameter of the well and frequency of data collection, monitoring wells can be equipped with dedicated pumps for water sample collection and/or pressure transducers for automated water level data collection.

5.2.12 Regulatory and Water Rights. Design considerations involving regulatory and water rights include volume, rate, and damage to senior appropriators imposed by law. State-issued permits usually dictate the allowable rates and volumes that can be diverted to a MAR facility. Damage to senior appropriators is usually defined in terms of water levels or water quality changes to affected water users. Various levels of monitoring and reporting may be required, depending on project size. Compliance is usually stipulated via agreements, contacts, permits, or legal decrees. Where catchment or basin water management plans are in place, various institutional arrangements can be used to maximize the benefits of MAR facilities and reduce adverse effects of multiple MAR facilities.

The Tamarack Ranch Recharge Project in Colorado, Wyoming, and Nebraska (see Case Study 4 in Appendix D) (Figure 5-2) is an example of how water rights hurdles were overcome to preserve critical habitat for endangered bird and fish species (piping plover, pallid sturgeon, whooping crane, and interior least tern). The project involves multiple federal and state agencies and multiple junior and senior water rights and court cases. The Tamarack I and II MAR projects include 11 recharge ponds that are fed by 16 wells, which serve to augment flows in the Platte River.

5.3 DESIGN CONSIDERATIONS FOR SUBSURFACE FACILITIES

The objective of subsurface facilities is to provide the most efficient connection between the recharge facility and the target aquifer. Subsurface facilities require the use of drilling equipment and drilling methodologies to provide access to the target aquifer. Obtaining a good connection with the target aquifer involves minimizing formation damage caused by drilling and

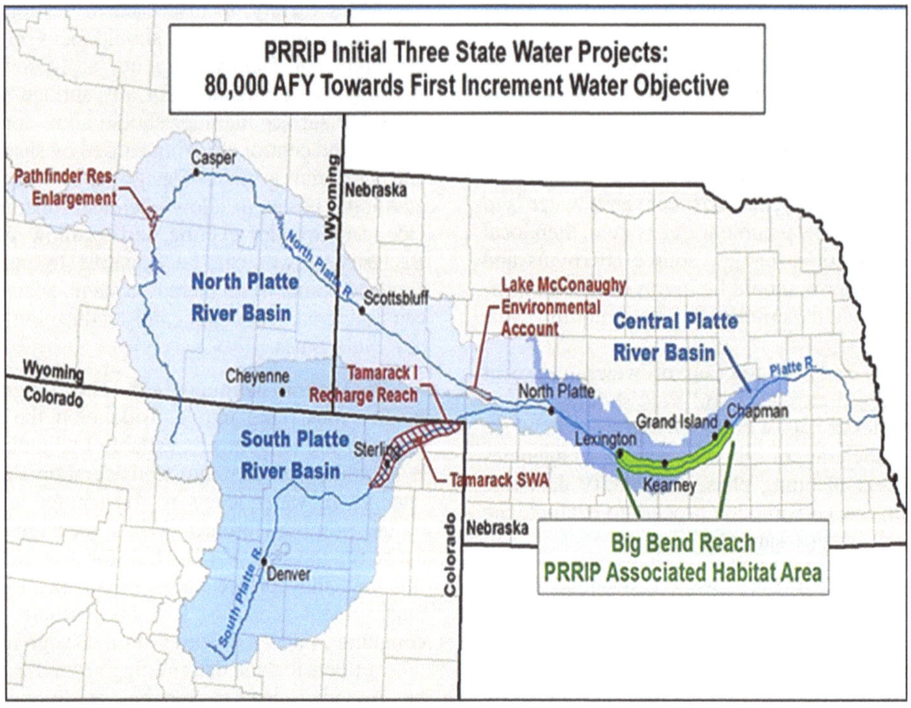

Figure 5-2. Tamarack Ranch recharge project.
Source: Altenhofen (2012).

minimizing the buildup of wall cake from drilling fluids and drill cuttings. As a result, adherence to detailed and appropriate drilling specifications, including properties of drilling fluids, and thorough well development are critical steps in successful well construction. Well design elements, such as the selection of the storage aquifer or interval(s), choice of casing and screen materials, their downhole configuration, filter pack selection, pump selection, and downhole flow control, can affect both the operation and the ultimate useful life of the facility. All recharge facilities have clogging issues, and this is especially true for recharge wells. Because of this, recharge wells and ASR wells may require more frequent redevelopment, and therefore, the design of these facilities should take this requirement into account.

General design considerations for subsurface facilities fall into the following categories and are discussed separately:

- Hydrogeology,
- Well field layout,
- Well design,
- Source water,
- Water quality and compatibility,
- Water treatment,
- Well pumps and piping,
- Corrosion protection,
- Flow measurement,
- Maintenance,
- Mound management,
- Groundwater monitoring, and
- Regulatory and water rights.

5.3.1 Hydrogeology. A thorough understanding of the geology, geochemistry, and hydrology of underground formations is a prerequisite to obtaining high recharge rates and volumes and for successful well or well field design. The geologic structure of an aquifer determines the size of the underground storage space. The unsaturated space (affected subsurface area multiplied by the soil porosity) is an upper bound of how much water can be stored in an unconfined aquifer. For confined aquifers, the upper bound is constrained primarily by lateral movement of the stored water under the natural gradient in the aquifer and by the ability to recover the water before it moves too far from the well. For both types of aquifers, the aquifer hydraulic conductivity and height of recharge mound determine how quickly the recharged water can move away in response to local gradients until encountering an aquifer boundary.

Sufficient effort needs to be spent in the planning and feasibility stages (Chapter 4) to collect, analyze, and evaluate how sensitive the aquifer configuration and aquifer parameters are to the various design considerations. Because of the wide range of hydrogeologic characteristics, there is risk in overestimating subsurface MAR capacity. This risk is why subsurface MAR facility design should include aquifer testing, pilot studies, and groundwater modeling, when possible.

Aquifer testing is used to measure hydraulic flow and storage parameters in a target recharge aquifer and to assess the well and aquifer for limits on recharge rates. The standard practice described in ASTM D5786 requires constant-rate pumping to produce measurable drawdown in both the pumping well and, ideally, in one or more nearby observation wells. Drawdown measurements collected during pumping are then analyzed to determine aquifer transmissivity, leakance, and storage coefficient values. For recharge wells, these parameters are used to estimate recharge rates and the extent and height of groundwater mounding. A mounding analysis is often required by regulatory agencies to assess potential impacts on subsurface structures, nearby open-pit mining operations such as sand and gravel mines, and existing wells in the area.

Throughout, and especially at the end of the constant-rate test, water quality samples should be collected and analyzed to establish the ambient water chemistry and to assess whether key parameters change during pumping. A step-drawdown test is conducted to estimate well yield and efficiency. A step-recharge test can sometimes be conducted to assess the aquifer's response to injection, which is usually different from its response to pumping. Like the step-drawdown test, water is injected at increasing rates for short intervals of 2 to 4 hours each. Three steps that bracket the expected recharge rate should be completed (Pyne 2005). The results of the step-recharge test are useful for establishing a baseline aquifer response that can then be compared with future tests to evaluate well clogging. In addition, the results of the step-recharge test are used to select a sustainable recharge rate for the well. Because the viscosity of water is sensitive to temperature, it is important to consider the potential effects of water temperature when evaluating recharge test results. Before injecting any water in a well, the recharge source water should be pumped to waste to remove any suspended sediments that could damage the aquifer at the borehole interface. This action is important because suspended sediments can be flushed from piping into the recharge water through reversal of flow in the pipe. A step-recharge test should not be conducted if earlier geochemical analysis has indicated that well conditioning is first needed before recharge to stabilize downhole swelling clays. Such well conditioning should be completed before any recharge into the well.

Upon completion of well construction and aquifer testing, there is typically a several month period during which facilities are designed, permitted, and constructed for equipping the well with a pump, motor, and other downhole facilities and for providing necessary wellhead facilities. During this period, interim recharge can be implemented, using a combination of temporary and permanent piping, valves, and other facilities to recharge an aquifer at a relatively low but steady rate. Initiating recharge on an interim basis can result in a substantial storage volume underground before completing construction of all facilities. No pump is in the well, so backflushing to remove clogging is not possible. Interim recharge cannot be conducted unless water of suitable quality is available and is considered to be geochemically acceptable for recharge.

5.3.2 Well Field Layout. Drilling plans and specifications should be prepared in the design phase. After drilling and testing, site-specific aquifer parameters are used to finalize the design of the downhole recharge and pumping equipment. Preliminary design is often based on local experience with wells in the storage aquifer and associated yields. If pipelines are being constructed between MAR wells, final pipeline infrastructure should be designed and constructed after MAR wells have been drilled and tested.

For facilities using multiple recharge wells, the project design requires carefully considering the aquifer location, extent, and depth; the anticipated drawup (mounding); and interference between wells. Preliminary locations of well sites should be evaluated in the feasibility stage to determine if they have favorable hydrogeological characteristics, are an adequate size and grade, and allow access for drill rigs. In the feasibility phase, multiple well site options should be compared from hydrological and cost (acquisition, easement, and infrastructure) perspectives (see Chapters 4 and 9 for more discussion).

Site preparation for recharge, recovery, or dual-purpose recharge and recovery wells (also known as aquifer storage recovery, or ASR wells) includes obtaining access to an appropriately sized property and obtaining necessary permits for the drilling of a well on that property and the discharge of water during testing and backflushing operations. The property should be large enough to accommodate a drill rig of sufficient size to construct the well or wells, including all of the ancillary drilling equipment, such as a mud system, compressors, roll-off bins for cuttings, and other equipment. Consideration should be given to future maintenance activities and access to the well or wells by equipment needed to maintain them, such as a pump rig with a tall mast or a drill rig if the well needs to be rehabilitated. In addition, there should be sufficient permanent space to store column piping and include facilities and equipment needed to address well clogging. Site layout design also may include identifying areas that should not be disturbed during construction, such as areas with wetlands, sensitive vegetation or habitat, and areas where access should be restricted. Provision for a future replacement well on the same site is also advisable.

5.3.3 Well Design. The proper design of wells used for MAR operations should consider aquifer properties, recharge and recovery rates, downhole equipment size, construction materials, and drilling method.

Existing wells should be considered for use in MAR operations. Advantages of using existing wells include the cost savings from not having to drill and construct a new well; to provide power for the pumping equipment; and to design, construct, and install the piping and appurtenances from the well to the treatment or distribution systems. Disadvantages of using existing wells could include the well not being in an optimal location for recharging the aquifer; geochemical incompatibility of the ambient water and the source water used for recharge; physical integrity of the well and particularly its screen for recharge; increased corrosion of the well casing because of the water used for recharge (particularly with older wells using mild steel); efficiency of the well; insufficient diameter to accommodate ASR downhole equipment; inappropriate storage intervals; and lack of access for retrofitting the well with additional distribution pipes, valving, and monitors. There may be other advantages and disadvantages associated with retrofitting a given well for recharge, but the potential cost savings warrant due consideration as part of the feasibility and design processes.

There is usually only one opportunity to demonstrate ASR viability at a site. If failure occurs because of shortcomings of the existing production well converted to ASR, there is often no second chance. If risk of failure is to be minimized, constructing a new well designed for efficient ASR operation is more likely to achieve that goal. The design and equipping of ASR wells and recharge wells is essentially the same, yet it is different from the design and equipping of production wells. Understanding the differences is important for an effective design.

The cost of a recharge well increases with increased casing diameter. However, the recharge rate also increases with increased casing diameter, so the cost-effectiveness of the well casing diameter needs to be considered carefully. The cost of a recharge well per meter of depth usually increases with well depth. The depth and screened intervals of the well relate to the depth of the groundwater basin and the aquifer(s) or storage intervals within an aquifer to be recharged. In addition, if multiple aquifers are to be recharged, the water levels or potentiometric heads in the various aquifers can be important considerations for which aquifers should be screened. For example, it might be necessary to install blank casing through a shallow, higher head aquifer to focus recharge on a deeper, lower head aquifer and avoid competing with cross-flow between the aquifers. Screen intervals set too shallow may not be suitable for intensive pumping, which is often implemented to backflush ASR wells, and may therefore be more susceptible to well clogging.

For dual-purpose ASR wells, the selection and thickness of screened or open-hole interval depends on the allowable degree of mixing between stored and native waters or on the possible occurrence of minerals in some producing intervals that could cause chemical problems. Where mixing is to be limited because of significant water quality differences, careful selection of the storage interval is necessary. Intervals suitable for production wells may not be the best intervals for ASR wells.

Well casing and screen materials commonly include mild steel, 304 and 316 stainless steel, and PVC, although other metals and inert materials such as fiberglass can be used if they meet the design criteria. Although the basic principles of design are similar for production and recharge wells, successful operation requires an understanding of important differences in the design of wells, wellheads, and well fields for recharge instead of for production or observation purposes. For both single-purpose recharge wells and ASR wells, experience has shown that periodic rehabilitation and redevelopment help maintain recharge rates. This maintenance is done most easily with pumps, so both types of well should be designed to provide ample room for pumping or redevelopment equipment.

The drilling method and associated fluid program can be critical to constructing an efficient MAR well. Care should be used when designing a drilling program, especially when working in lower permeability formations. Commonly used bentonite-based drilling fluids can block pore spaces and can be very difficult to remove. As with any well, proper well development methods should be written into the construction specifications by a qualified hydrogeologist or drilling engineer.

The construction of recharge wells is similar to that of groundwater production wells; however, some design elements may differ. Drilling equipment and techniques have expanded and improved over the years such that there is now a wide variety of drilling options available. The proper choice of drilling method should consider all the following:

- Ultimate depth of the wells in the well field;
- Well design, including the ultimate diameter of the well, the appropriate dimension of the annular space between the casing and the borehole wall, and the space needed between the pump column flanges or couplings and the inner casing so that there is adequate room for downhole tubing and equipment;
- Lithology of the subsurface materials, including the presence of clay, volcanics, hard materials such as rock or caliche, and/or large river cobbles or clasts;
- Cost considerations, including the scale and budget for the project; and
- Schedule constraints, including when the well or facility is needed.

Table 5-1 lists several commonly used drilling techniques and their advantages and disadvantages with respect to drilling wells for MAR projects. A more complete description of drilling techniques can be found in Driscoll (1986), Roscoe Moss Company (1990), and in ASTM Standards D5781, D5782, and D5875, among others.

When constructing a recharge well, it is best to choose a drilling method that minimizes damage to the formation and the buildup of wall cake on the borehole wall; both can limit the flow of water across the borehole–well interface. For example, mud

Table 5-1. Commonly Used Drilling Techniques.

Drilling Method	Description	Advantages	Disadvantages
Cable tool	Percussion drilling using a heavy steel bit suspended from a cable that is raised and dropped to advance the hole. Cuttings are removed by a bailer on a wireline; casing may be driven to prevent sloughing in unstable formations.	Low cost; produces straight holes; wide range of diameters possible; casing can be driven to secure unconsolidated formations; drilling fluids not usually necessary except in dry formations; minimal skin damage to formation; can obtain representative water samples.	Very slow drilling technique, especially at large depths; flowing sand formations can pose a significant challenge. Difficult to place good seal between casing and formation. Can be a problem if recharging under surface pressure.
Bucket auger	A large-diameter bucket auger is rotated to advance the hole and then extracted to remove cuttings.	Large diameters [0.6 to 1.2 m (2 to 4 ft)] can be drilled quickly in unconsolidated formations; relatively inexpensive; no drilling fluids needed.	Depth limited to less than about 30 m (100 ft); only good in unconsolidated formations.
Conventional mud rotary	A drill bit is rotated, and drilling fluids are circulated down the pipe connected to the bit. Fluids and cuttings return between the borehole wall and the drill pipe. Drilling fluid can include additives to increase density and viscosity to carry cuttings to the surface.	Can penetrate alluvium and/or rock to great depths; relatively quick, cost-effective method; can drill through relatively unstable formations as long as drilling mud is well managed.	Formation samples can be poor because of mixing in hole; water samples are difficult to obtain; lost circulation can be a problem; can generate wall cake and formation damage that interferes with fluid flow into or out of well; well development can be difficult; and hole deviation from vertical is common.
Reverse air circulation	Drilling fluids are circulated down the hole between the borehole wall and the drill pipe, then return with cuttings through either the drill bit or through a reversing substitute joint ("sub") some distance above the drill bit, and out through the drill pipe. Can be done with pressurized air (see air rotary following).	Minimal formation damage as compared with conventional mud rotary drilling; provides both good formation samples and representative water samples.	Risk of lost circulation; hole deviation from vertical is common; and it is more expensive than mud rotary. Best if have formation capable of maintaining an open hole.
Air rotary	Instead of circulating fluids, pressurized air is used to lift cuttings from the borehole. Drill fluids such as water or foam can be added to lift cuttings to control dust; one or more air compressors may be required, depending on the depth and diameter of the borehole.	Good for obtaining formation and water samples; can readily estimate formation productivity through observation of the discharge flow rate.	The depth of penetration using reverse air drilling can be limited in highly productive formations or at great depth below the water table exceeding the capacity of the air compressor(s). Often must switch to flooded reverse or conventional rotary drilling.
Flooded reverse	Using the reverse circulation method, a fluid level is maintained in the borehole, usually above the static water level to provide positive pressure to the borehole wall. The fluid returns through a reversing substitute joint ("sub") above the drill bit carrying drill cuttings up the hole via the drill pipe.	Provides borehole stability in loose formations or in fractured rock with minimal formation damage.	Mud cake on borehole wall can still be a problem if drilling mud additives are not well maintained or insufficiently developed out of the hole. Can be more expensive than other methods.

rotary drilling is an excellent, relatively inexpensive drilling method. However, if not properly managed, this method can cause both formation damage and wall cake buildup (which can lead to a zone of high head loss also known as the skin effect). Using reverse air drilling can reduce formation damage and, if no additives are used, produces the least amount of wall cake of any of the drilling methods listed. Reverse air drilling may not be suitable if the formation is loose or unstable, or if the borehole has a large diameter. An experienced hydrogeologist or well driller should be consulted before final selection of the drilling method.

Once a drilling method has been selected, a registered or professional geologist or engineer should prepare a set of technical specifications for well drilling and construction. The technical specifications detail the methods and materials acceptable for drilling and construction of the well and provide clear guidance to the contractor regarding his or her responsibilities, how adverse conditions are to be handled (such as slow drilling), and the basis for payment (including hourly rates, footage rates, or unit costs). The technical specifications and associated bid table become the basis for development of bids by qualified drillers; however, the selection of a drilling contractor is typically based on the lowest cost and the most qualified bidder.

5.3.4 Source Water. The source water used for subsurface recharge may include treated or untreated lake or river water from the same watershed, imported water from another watershed, treated potable water from either surface or groundwater sources, treated urban stormwater, or reclaimed wastewater from sewage treatment plants or industrial operations such as coal seam gas production. Groundwater from an overlying or underlying aquifer may also be used for storage and from the same aquifer at a different location, such as where water quality may be better. These sources may be available separately or used in combination. Each recharge well may have its own water supply, or there may be a common water supply system for several wells. One component of the supply system would be a facility for creating the necessary pressure required to maintain the flow of the supply water and provide the required recharge pressure. This facility would either be a booster pumping facility or a pressure reduction facility, depending on the base water supply. An isolation valve should be installed at each well. This valve enables a well to be taken out of service without having to shut down the entire system.

5.3.5 Water Quality and Compatibility. Source water can vary widely in quality but must be compatible with target aquifer quality to avoid precipitation of minerals that can cause clogging of the well screen, gravel pack, or aquifer pore space near the well. Similarly, solutions of minerals in the aquifer can mobilize fine materials that can clog a well and/or reduce the pH, causing further damage to minerals in the aquifer close to the well. Compatibility of source water and native groundwater should be identified in the planning phase and is a key criterion in the design phase. Water quality includes temperature differences between source water and the aquifer. Viscosity increases as temperature decreases, and injecting water with temperature lower than the target aquifer can result in a reduction of recharge rates compared with well yields measured in pumping tests.

The storage volume for a MAR project should also take into consideration that a buffer zone is created at the outer edges of the recharged water. The buffer zone separates the recharged water from the groundwater in the aquifer and is usually not recovered. The volume of water in the buffer zone reduces the volume of usable water calculated beneath the project site, with the reduction depending on the quality of the extracted water and its intended uses.

5.3.6 Water Treatment. All water introduced to MAR recharge or ASR wells should be as free of suspended solids and entrained air as possible. Suspended solids may come from flow reversals in piping bringing water to the wellhead, pieces of cement from cement-lined ductile iron pipes, rust in mild steel well casings, and many other potential sources. Including a filter at the wellhead is often a reasonable precaution. It is a lot easier to keep solids out of a well than to remove the solids once they are in the well.

Entrained air can rapidly clog a well in a few minutes. The design and equipping of the well and the wellhead facilities, and their operation, should ensure that air does not get entrained in the recharge water. This design usually entails use of air and/or vacuum relief valves on the wellhead casing annulus and column pipe. Any entrained air is carried down the well and out into the formation, where it can air-bind the aquifer. Eventually the entrained air goes into solution under the increased pressure downhole; however, in the meantime, the well may rapidly clog. The timing and sequencing of start-up and shutdown of recharge operations need to be managed carefully to avoid entraining air.

Growth of bacteria in and near the well screen contributes to clogging and is best controlled by maintaining a small disinfectant residual in the well at all times including extended storage periods exceeding about one week. A trickle flow of drinking water containing a small chlorine residual is often used, displacing the well volume about once a day. Many recharge water sources contain dissolved organic carbon, nitrogen, and phosphorus, all of which encourage downhole microbial growth. As discussed in Section 3.8, research has documented the attenuation of disinfection by-products during aquifer storage, particularly in anoxic conditions. Dechlorination of the recharge water to meet regulatory criteria for disinfection by-products is therefore not recommended.

The design of a facility needs to take into consideration any pretreatment that may be necessary to minimize clogging of the well. For example, the sodium adsorption ratio may influence clay dispersion and clogging near the well, especially when fresher water is introduced to a brackish aquifer containing sodium montmorillonite clays. Alkalinity and pH adjustment of the recharge water may be needed to stabilize dissolution of minerals in the aquifer that would otherwise clog the formation with fine materials. Calcium carbonate in the recharge water may precipitate on well screens and around pumps during extended recharge periods. Ozonation and advanced oxidation treatment processes can yield recharge water with exceedingly high oxygen concentrations, which can be aggressive to minerals in the aquifer if the water is not adequately stabilized. A geochemical analysis of the recharge water quality, ambient groundwater quality, and the minerals in the storage aquifer can identify potential issues during the feasibility study and pilot testing phases, before they become operational issues. Measures can then be integrated into the design and operation to address these concerns.

Water treatment facilities might include disinfection, dechlorination, pH adjustment, or filters to remove particulates from the source water before recharge. Projects with water sources in which turbidity and TSS concentrations vary over a large range may need to have water quality thresholds established so as to prevent the diversion of water that could cause excessive clogging. Developing these thresholds requires a good understanding of how TSS loading affects facility performance, the cost of cleaning the facilities, the backflushing frequency and duration for recharge wells, and the value of the water lost or gained.

For wells used for aquifer recharge of surface water, the removal of suspended solids can potentially be achieved through the use of large-volume settling basins, often in conjunction with the addition of flocculants or coagulants. The addition of flocculants or coagulants, which include a wide range of chemicals such as alum, ferric chloride, lime, and organic polymers, enhances the settling rate and reduces the basin volume necessary for adequate detention. A United Nations Report (UN 1975) notes that water detained for 2 to 3 days at a constant 5 m (16.4 ft) water depth reduced the suspended solid load by 300 to 1,000 mg/L, while the proper dose of chemical coagulant achieved the same results in a few hours. Miller (1980) found that either alum or iron sulfate at 2 to 40 mg/L was effective for turbid surface waters and that 40 to 300 mg/L was required for effluent from wastewater treatment operations using sedimentation and activated sludge treatments. Extended periods of time are required to settle suspended particles without the aid of coagulants. However, any coagulant residuals that are in the recharge water may cause clogging of a surface recharge facility or recharge well, so flocculants should be considered only after other options have been explored to reduce suspended solids. Even with flocculants added, filtration of the settled water is usually needed to further reduce particulates in the water before recharge through a well. Adjustment of pH and disinfection may also be necessary, except perhaps for recharge into karst limestone aquifers with very high transmissivities.

Recharge water quality criteria are established and enforced by the USEPA underground injection control (UIC) program or its local delegate, typically a state agency, to ensure protection of underground sources of drinking water (USDWs). Very few MAR projects in the United States recharge partially treated surface water into storage aquifers through wells. However, excess water from stormwater runoff and high lake levels has been used to recharge karst limestone aquifers and other similar aquifers with high secondary porosity and permeability through what are referred to as drainage wells (Sheffield et al. 1995). This type of well was fairly common in central Florida, where it was estimated that there are more than 400 drainage wells in Orange County (Taylor 1993). Permitting of drainage wells was prohibited by USEPA from 1983 to 2013, so most drainage wells in the United States are many decades old. More recently, site-specific approval for such wells has been delegated by USEPA to the states, opening an opportunity for managed aquifer recharge with appropriately pretreated surface water (USEPA 2013).

Post-recovery water treatment can include disinfection, filtering, pH adjustment, membrane treatment, and/or softening if the recovered water is anticipated to be hard. If recovered water is to be used as a source of supply to a potable water system, then local agencies that have jurisdiction over new source approvals and regulate drinking water systems are involved in establishing post-recovery water quality criteria and determining treatment requirements. If recharge has been through ASR wells instead of recharge wells, the only treatment typically required after recovery is to restore the disinfectant residual. For MAR with recharge wells, the recharge water mixes with the ambient groundwater in the aquifer, gaining water quality constituent concentrations through dissolution of minerals as the water moves through the aquifer to the point of withdrawal.

The selection, design, and construction of water treatment facilities require specialized expertise and, thus, should be performed by licensed professionals with extensive experience in water treatment. It is prudent to consult with regulatory agencies early in the MAR project planning phase to identify data that should be collected during the well construction process to evaluate the need for, and type(s) of, water treatment. All treatment processes should be integrated into the overall MAR design.

5.3.7 Well Pumps and Piping. Well pumps and piping need to be selected with the well capacity, desired range of recharge rates, and recharge and recovery water levels in mind. Well pumps for most ASR wells fall into two general categories: vertical line-shaft pumps or submersible pumps. Vertical line-shaft pumps use an aboveground motor and a long line shaft to turn the pump impellers deep in the well. Submersible pumps are vertical pumps that use a submerged motor that is mounted below the pump assembly to turn the pump impellers.

Vertical line-shaft pumps have several advantages for ASR wells over submersible pumps. An important advantage is that it is possible to recharge directly backward through a line-shaft pump and use the pump column and bowls as recharge equipment. In aquifers with deep water levels, this method can be used to eliminate potential air entrainment during recharge caused by cascading of recharge water. In aquifers with water levels that rise quickly toward the ground surface during recharge, it is usually best to recharge in the annular space around the pump using what is known as annular recharge. Using this method requires sealing the discharge head base to prevent leakage when the well goes under pressure. Either a fabricated steel discharge or a cast iron head can be machined flat to mate with a steel sole plate, using either a gasket or O-ring to seal the surface.

When recharging through a vertical line-shaft pump, the pump bowls work like an orifice plate to restrict recharge flows and maintain a positive pressure throughout the pump column and at the ground surface. The disadvantage to this method is that the range of flows needed to maintain positive pressure is usually narrow, and recharge flow rates cannot be changed much. If the recharge flow rates are reduced to the point where it is not possible to have pressure in the recharge piping at the ground surface, a vacuum can occur and air can be pulled into the recharge water through either the line-shaft stuffing box or seal, or the threaded column joints, which in some cases can be straight threaded and not completely sealed.

Vertical line-shaft pumps have a line shaft throughout the pump column that runs through bearings. The line-shaft bearings require lubrication, and the way they are lubricated defines the type of vertical line-shaft pump. Vertical line-shaft pumps can be of open line-shaft construction, where the pumped water lubricates the bearings, or of enclosed line-shaft construction, where a separate fluid, either water or oil, is used to lubricate the bearings. The importance of this difference to MAR wells is that if oil is used as the lubricating medium, oil is introduced to the well and ultimately floats on the water surface in the well. This method can provide a means for oil to enter the recharge flows and to be pushed into the aquifer. Oil-lubricated line-shaft pumps should be avoided in MAR wells.

Submersible well pumps are typically used in MAR wells with deep pumping water levels, where noise at the ground surface is an issue, or where wells with poor vertical alignment do not allow proper operation of a line-shaft pump. Vertical line-shaft pumps are common with pumping water levels as deep as 150 m (500 ft). Vertical line-shaft pumps can be used with pumping water levels up to 300 m (1,000 ft) with careful pump selection and have been used with pumping water levels up to 450 m (1,500 ft) using special bowls with substantial lateral clearance for the pump impellors. But column and shaft stretch caused by pump thrust limits the total depths of vertical line-shaft pumps.

Most submersible pumps do not allow water to flow backward through the pump because this practice can damage the thrust bearing in the pump motor. Oil-filled submersible motors are sometimes the exception to this; however, a pump expert must be consulted if this application is desired. This choice often requires a separate recharge tube in the well, which sometimes increases the required well diameter, or a special downhole valve to restrict and divert recharge flows around the submersible pump, which can result in increased downhole equipment costs.

Pump columns can be either threaded and coupled or flanged on MAR wells. Sometimes the designer selects a coated column, in which case flanged joints should be used. The use of a coated column is not typically recommended unless absolutely required because small breaches in the coating can result in pinhole corrosion, which can accelerate column failure. Flanged column pipe, such as for epoxy-coated or stainless-steel pump columns, typically requires larger casing diameters than does threaded column pipe.

MAR facility electrical power is the same as for a typical well facility. Power is typically low or medium voltage, usually at the utility's preference. Deep, high-horsepower submersible pumps may require medium voltage power to minimize drop cable sizing. If variable frequency drives (VFDs) are used in the power system, it is better to use a low-voltage VFD and then boost the voltage to medium voltage after the drive. As with a typical well system, it is important to match the pump motor windings to the VFD application. With submersible pump motors, a filter is often required on the output of the drive to prevent damage to the motor windings.

Wellhead piping extends from the connection to the source water to the discharge in the ASR or recharge well. The discharge is for handling of backflush flows and for recovery of the stored water. Pipe sizes are designed using normal velocity and friction loss calculations. Wellhead piping systems depend on the recharge system selected. When recharging through a vertical line-shaft pump, a single bidirectional flowmeter can be used, and an injection loop around the above-grade check valve can be installed. Typical bidirectional flowmeters are magnetic meters, although ultrasonic meters have also been successfully used. Piping to discharge periodic backflush flows is also required, and many ASR systems provide a second, separate flowmeter for these flows. This meter provides a simple, reliable method for reporting discharge flows that does not involve additional programming that may be required if only one meter is used for recharge, recovery, and backflush flows.

The initiation of recharge in MAR systems sometimes requires starting recharge flows from a high system pressure to little or no pressure on the downstream side of the recharge valve. Careful selection of a valve capable of starting these flows with little or no cavitation damage is required. In addition, if the piping design requires throttling of recharge flows, proper valve selection is critical to the life of a MAR facility. Flow regulation can be accomplished using both butterfly and gate valves, although regulation of very low flow rates with butterfly valves may be more difficult than with globe valves. Downhole control valves are frequently used to prevent entrainment of air during recharge. Downhole flow control valves are installed above the submersible pumps to restrict flow at depth and thereby prevent cascading water and entrained air. They have been installed above or below the pumps for vertical line-shaft turbine pumps. Lay-flat tubing has also been used successfully to prevent air entrainment in shallower wells.

The city of Phoenix, Arizona, uses the reverse siphon method, which eliminates the need for a downhole flow control valve in one of its ASR wells. The reverse siphon requires sophisticated automated valving systems but is advantageous because there are no downhole control mechanisms, thus allowing operators to observe and maintain all mechanical equipment associated with well activity above ground (see Case Study 9 in Appendix D for more details). Considerable care is needed with this operational approach to avoid inadvertent air entrainment or damage to the screen formation pack that could cause sand inflow to the well.

5.3.8 Corrosion Protection. Most large-diameter recharge wells are constructed with casing and screen of steel, or an alloy containing steel. Once the well is completed, the exterior of the casing and screen are inaccessible for inspection. Whereas the interior can be inspected by downhole cameras, the procedure is costly and requires removal of the pump to inspect the well interior in its entirety. Therefore, the potential for corrosion of the well should be carefully considered when the well is designed.

The factors to be considered that indicate corrosive waters are the following (ASCE 2001):

- Low pH, indicating acidic water;
- Dissolved oxygen (DO) in water;
- Hydrogen sulfide in water;
- High total dissolved solids (TDS more than 500 mg/L);
- Carbon dioxide higher than 50 mg/L;
- Chloride ions higher than 300 mg/L; and
- High temperature of water in well.

If any of these factors are common in the area where the well is to be constructed, mitigation measures such as corrosion-resistant steels and nonmetallic casing and screens should be considered. Greater wall thickness of metallic casing and screens to provide sacrificial metal, coating of metallic surfaces, and cathodic protection systems are also options.

An option is to construct the well with PVC casing, when possible. This method is frequently less expensive and eliminates the problem of well clogging caused by corrosion products from the steel casing. More importantly, it reduces or eliminates the problem of handling the first flush of rusty, colored water when a recharge or pumping well is pumped after redevelopment, or when an ASR well changes from recharge to recovery.

Roscoe Moss Company (1982, 1985, 1990) states that if one designs a municipal well for 50 or more years of operation, the use of stainless steel materials, whose corrosion rate is negligible in potable water, is mandated in most areas of the United States and elsewhere, where waters are generally alkaline. In addition, doubling the wall thickness will extend its life four or more times, except in severely corrosive waters.

5.3.9 Flow Measurement. Metering should measure both recharge and, for ASR wells, discharge. A flowmeter should also be placed on the line used for waste discharge during backflushing operations to evaluate overall recharge efficiency. If a trickle flow line is included as part of the well design, thereby maintaining a disinfectant residual downhole during extended periods of no recharge or recovery flow, then a simple flowmeter should be installed on the trickle flow line. Flowmeters should record instantaneous rates and cumulative flow. Almost any type of reliable flow-rate measuring device for a pressure water system is satisfactory, with the particular choice probably dependent upon the specific MAR project. Many ASR wells use bidirectional magnetic flowmeters. The Los Angeles County, California, Flood Control District uses differential pressure devices consisting of flow tubes and orifice plates to monitor the flow rates of 210 recharge wells at its seawater intrusion barrier projects. Portable flow gauges are used rather than dedicated gauges when cumulative flow volumes from individual wells are not required.

5.3.10 Maintenance. An operations and maintenance program should be designed for any MAR project, as discussed in Sections 5.5.2 and 7.2. A key focus of operations and maintenance is maintaining the recharge capacities of individual MAR wells.

Clogging is inevitable for any subsurface MAR facility. Clogging can be caused by several factors, including suspended sediment entering the well from the source water, chemical and mineral precipitation, and biological growth. Wellhead filtration can be designed at the wellhead if source water solids are high. Wellhead filtration may include in-line or bag filters, or sand filters. The degree of filtration needed depends on the nature of the clogging material, the permeability and porosity of the aquifer, and the well design. Filtration design should be based on sampling of source water and aquifer matrix and, if needed, on laboratory column testing. Pilot test design should rely on column testing or modeling to gain insight into the various clogging processes that may affect well performance. The pilot test can then be designed to gather sufficient measurements of water quality parameters and rates of clogging so that results can be reliably interpreted, and preventive measures can be evaluated. Section 5.4 provides more discussion on pilot testing design.

Johnson (1981) reported the principal causes of clogging in a recharge well to be the following:

- Gas binding, or air entrainment, in the aquifer;
- Suspended particles in the recharge water;
- Bacterial contamination of the aquifer by the recharge water and subsequent clogging by bacterial growth;
- Chemical reactions between the groundwater and recharged water, causing precipitation of insoluble products;
- Ion-exchange reactions that could result in clay-particle dispersal and/or swelling;
- Precipitation of iron in the recharge water as a result of aeration;
- Biological changes in the recharge water as well as the groundwater;
- Swelling of clay colloids in the dewatered portion of the aquifer; and
- Mechanical jamming of the aquifer materials caused by particle rearrangement when the direction of water movement through the well and aquifer are reversed.

Clogging material that accumulates in the well needs to be purged via periodic backflushing. The frequency and duration of backwash cycles should be determined during the pilot study phase and considered in the overall recharge efficiency target for the site and project. It is important that wells be backflushed at appropriate intervals to avoid irreversible clogging. Design options include automatic backflushing using programmable logic control (PLC). Besides PLC, an automatic backflush system requires automatic flow control valves and a water-level pressure transducer to measure injection drawup pressures over time. The backflushed water is usually discharged to waste. Backflushing can also be necessary after a well has been idle for some time. Therefore, routine backflushing and management of the resulting purge water and sediment should be included in standard operation practices.

Backflushing frequency may vary from once every few days to once every few weeks. For recharge into highly transmissive aquifers, such as karst limestone, backflushing may be unnecessary, other than occasional well redevelopment every few years to remove accumulated solids that may be present in the recharge water. For ASR wells storing drinking water in karst limestone aquifers, purging water to waste for an hour at the beginning of a seasonal recovery period may be sufficient. Backflushing procedures also vary. A single backflush for an hour may be sufficient for PVC-cased wells storing drinking water. For mild steel cased wells, a few hours may be needed to purge rust from the well. Conducting two or three backflush cycles in sequence may be needed to restore well performance, restoring filter pack development to accommodate bidirectional flow. An extended recovery period is usually most effective. That is partly why ASR wells tend to have more sustained hydraulic performance than recharge wells.

5.3.11 Mound Management. The design of a recharge well must take into consideration the anticipated recharge rate and the amount of available head to work with. It should also take into consideration seasonal or long-term changes in water levels in the aquifer in the general vicinity of the proposed project. When recharge rates exceed aquifer capacity, then there is a potential for waterlogging soils or even flooding the surface facilities or nearby structures in the vicinity of the recharge well. In most cases, the pressure head placed on the aquifer should be less than the fracture pressure of the aquifer, unless the intent is to fracture the aquifer, in which case additional permitting and design may be required. Mound management scenarios can be evaluated using analytical or computer models (see Section 9.6).

Most recharge wells, and almost all ASR wells, store water in confined or semiconfined aquifers. During recharge, it is not uncommon for wellhead pressures to be above land surface, typically by up to about 138 kPa (20 psi). The wellhead therefore needs to be properly sealed to withstand such pressure. For large-diameter casings, the upward force on the wellhead flange can be substantial. For wellhead pressures exceeding about 206 kPa (30 psi), special measures may be needed to effectively seal the holes in the wellhead flange for transducer cables, trickle flow lines, power cables, and other tubing. Wellhead pressure should be sufficient to prevent air entrainment, such as at least 14 kPa (2.0 psi). With appropriate wellhead piping design, once pump column pressure reaches about 14 kPa (2.0 psi), recharge down the casing annulus can be added, thereby slightly increasing recharge flow rates. The increase in water level above the static water level during recharge should not exceed the drawdown available by pumping the well, preferably with a substantial safety factor to accommodate potential loss of specific capacity caused by well clogging. The pump in a recharge well or ASR well is usually set deeper than for a typical production well.

The specific injectivity (the rate of recharge per unit increase in water level) of a recharge well or ASR well is typically less than the specific capacity. For an unconsolidated (sand and gravel) confined aquifer, the ratio of specific injectivity to specific capacity [gal./min (gpm)/ft of water level rise or fall)] is typically within a range of 0.3 to 0.8, with a best estimate of 0.5 if no data or local experience is available. For a consolidated aquifer (such as limestone, dolomite, sandstone, or basalt), the ratio is typically about 0.8 to 1.0. Data defining this ratio are usually not available during design of wellhead and well-equipping facilities, so flexibility is needed to adjust operations to match actual performance after the well begins operation. Various hypotheses have been offered as to the reason(s) for this ratio. Further investigation is needed.

For ASR wells storing freshwater in confined, brackish aquifers, or other aquifers containing poor water quality, wells may be spaced closer together than for a typical production well field. Well interference during recovery periods is accepted, with the associated increase in power costs and reduction in individual well production rates. Recharge water creates a buffer zone that pushes brackish or other poor-quality water away initially from a recharge well, improving water quality progressively from brackish water to freshwater until the remaining water recharged, called the target storage volume, meets design criteria for recovery. Recharge water in the buffer zone is never recovered. However, once the target storage volume has been created, long-term recovery efficiency steadily approaches 100%. Subsequent operating cycles achieve 100% recovery efficiency during each cycle.

The volume comprising the buffer zone depends upon many factors and is usually expressed as a percentage of the target storage volume, typically ranging from 0% to 50%. Storage in a fresh aquifer containing ambient groundwater with excellent water quality and no significant geochemical issues might not need a buffer zone. Storage in a semiconfined, highly brackish aquifer, with potential geochemical issues, might need a 50% buffer zone. The value of the water forming the buffer zone is usually a relatively small component of the total cost for a well installation and is typically amortized as part of the well construction cost. This target storage volume (TSV) approach has been demonstrated to easily control arsenic mobilization and attenuation at many ASR sites, without the need for deoxygenation of the recharge water.

5.3.12 Groundwater Monitoring. As with surface recharge systems, monitoring is necessary to evaluate the aquifer's response to recharge compared with the design targets and to evaluate project operations. Monitoring the quality of source and recovered water is key to understanding water compatibility issues and future O&M requirements.

Monitoring wells may be required by regulators to verify that water levels do not rise to the ground surface and cause flooding or waterlogged soils and possible liquefaction. Nested monitoring wells should be located near recharge wells to evaluate vertical gradients and the aquifer response. Monitoring wells should be placed on the facility perimeter to monitor hydraulic gradients, flows, and water quality. Monitoring injection, recovery, and backflush volumes is necessary to determine overall recharge efficiency.

Monitoring wells are installed using the same drilling and construction techniques as those used for recharge wells. However, monitoring wells typically have smaller diameters and are completed with PVC casing. Monitoring wells should be drilled to the same depth and screened over the same zone(s) as the recharge well to enable collection of representative water level and water quality data (see ASTM Standards D5092 and D5521). Depending upon the depth and diameter of the well and frequency of data collection, monitoring wells can be equipped with dedicated pumps for water sample collection and/or pressure transducers for automated water level data collection.

Electrical conductivity and temperature are important to monitor and can be used as indicators for identifying recovered source water. Obtaining baseline data on these parameters, for both the source water and the target aquifer, is needed before initiating MAR operations. Ongoing monitoring of source water and recovered water is important to understand temporal variations in each and to gain an understanding of mixing within the aquifer. Turbidity monitoring has also been useful at some sites for the same reasons. Measurements of flow rate and head in the recharge well help identify variations in specific capacity and may be of use to determine when backflushing is needed.

A SCADA system should be used to collect data from all measuring devices for engineering analysis. Level, flow, and water quality data can be used to optimize the system and as the basis for O&M, design improvements, and phased expansion.

MAR facilities can be controlled manually, automatically, or semi-automatically. A prototype MAR surface recharge facility may be mostly manually controlled because these facilities are in place to test the MAR concept and identify not only if a MAR project is feasible in a particular location, but also to identify the best way to operate the MAR facility. Pilot testing on a small recharge basin is a normal approach. Lessons learned are then applied to the design and operation of expansion facilities.

For recharge wells or ASR wells, initial demonstration testing should be conducted on one or two full-size wells, not a small-diameter well. Results tend to be scale-dependent, so testing small gets small results. There is usually only a single opportunity to demonstrate success, so it is advisable to take every opportunity to incorporate design and operational features to achieve goals. Because wells in the same area usually have different production or recharge rates, it is helpful to have more than a single demonstration test well. Once success has been demonstrated on the initial test well(s), appropriate design and operational adjustments can be made for subsequent wells.

After the operational needs are identified from pilot or demonstration testing, some degree of automation is typically incorporated into the design of the full-scale or expanded MAR project. The level of automation applied is usually an owner and/or operator preference and can range from very simple to complex. In general, system automation types can be categorized in the following groups: monitoring only, semi-automatic, and fully automatic.

The next sections discuss what automation controls could be included in these groups.

Monitoring Only. Remote monitoring of a system's operation is typical of many MAR applications. Monitoring systems typically consist of monitoring instruments of the various system processes in a facility with both local signal readouts and the ability to combine and transmit these signals to a central location through radio or signal cables. The central location is typically furnished with an interface computer that receives signals for the various processes and displays them on various graphical screens. The interface computer can also process these signals, store them, and manipulate them so as to display different units of measurement, perform calculations on the different signals, and provide warnings or alarms if a programmed instruction or condition is not met.

For example, if recharge flow into an ASR well and the water level in the well are being monitored, the interface computer can be programmed to alarm the operator if the flow reaches a high or low value, or if the well-specific capacity falls below a certain value. This arrangement provides the operator with quick information that something may not be working properly so that he or she can investigate the problem.

In manual systems, it is typical to monitor and remotely transmit the following signals:

- Recharge and recovery flow rates,
- Recharge and recovery daily and cumulative volume totals,
- Flow rates in system piping at key locations,
- Wellhead pressure and pipeline pressures at key locations,
- Running status of all pumps, and
- Water levels in all wells and monitoring wells.

Many MAR systems operate in a remote monitoring mode because it is simple and allows the operator to see the entire system operation from a central location.

Semi-Automatic. A semi-automatic scheme takes the monitoring only scheme one step further and provides actions based on the observed signals and computer calculations. For example, recharge and recovery flows can be monitored and the operation terminated if those flows are outside a preprogrammed range. For example, if recharge flows drop to a preprogrammed level, recharge can be automatically terminated by the computer closing a motorized valve. Or if recovery water levels drop to a level that may damage a pump, the pump can be ordered to be shut down until an operator diagnoses the problem.

Semi-automatic operation can also include regular backflushing of the ASR wells. The system can be programmed to backflush at a set time interval or to backflush when specified water levels in the well are reached. Motor-operated valves are required to shut down recharge, open valves to waste, run the pump, and then initiate recharge after the backflush cycle is complete. Automatic backflush systems are in place at a few sites, but it is more common to shut down the well and manually backflush it after high well water levels are reached.

Semi-automatic actions are typically used to safeguard the system in case operational flows and pressure fall outside a predetermined range. Many ASR systems provide the semi-automatic safeguards in their systems to stop a process that could cause harm to the wells or equipment.

ASR wells and recharge wells equipped with pumps for backflushing typically require reliance upon SCADA controls for operation, including time delays and set points intended to not only protect the equipment but also to prevent air entrainment during recharge and backflushing operations. Operational adjustments may be needed to accommodate changing static water levels, changing specific capacity and specific injectivity, and variations in flow rate available for recharge. It may be difficult to keep track of all this with manual operation. Semi-automatic and automatic operation are generally needed. Otherwise, operators may tend to look for ways to simplify operations, such as by not backflushing when needed and not recharging except under ideal conditions when steady flows are expected to be available for a long period.

Fully Automatic. Fully automatic ASR systems exist; however, most systems are equipped to be semi-automatic. Fully automatic systems have been developed to perform complete operations at the push of a button, or to do complete operations at a scheduled time. For example, MAR systems have been designed so that an operator could completely control operations at multiple wells with the push of a button from a central computer. System operations of recharge, recovery, backflush, resumption of recharge after backflush, and even electrical power generation from the recharge flows going down the well were processes set up to occur at the command of an operator.

An example of a fully automatic system is the Dominguez Gap Barrier Project seawater barrier system in southern California. This system injects water along the Pacific coast into 33 wells for extraction further inland. The system was designed with an automatic air charge and airlift redevelopment system and a microfiltration system to treat and reuse the backflush water. The process was set up to automatically stop recharge at one well, automatically air charge and airlift that well for one day, collect the airlifted water in a tank, and treat the backflushed water for reuse the next day. One well was air charge and airlift developed every other day this way so that every well was automatically backflushed every 66 days.

In summary, some level of automation in MAR systems is common; a semi-automatic approach is the most common. The fully automatic systems generally become more complicated as substantial programming and sensitive instruments are required to be maintained. The more automation that is applied to a MAR system, the higher the level of technical expertise required of the operators. Depending on the geographical location, there is usually a point in system automation where operational costs may actually be higher than for a more semi-automatic system.

5.3.13 Regulatory and Water Rights. Design considerations involving regulatory compliance and water rights are the same for subsurface MAR projects as they are for surface MAR projects, and they include volume, rate, and damage to senior appropriators (see Section 5.2.12). However, water level effects are more likely to occur over a larger area in a confined aquifer than in an unconfined aquifer because of the lower aquifer storage coefficient in confined aquifers. The recharged water may extend only a few tens of meters (hundreds of feet) from a recharge well, but the pressure effects may extend for several miles (kilometers). In Colorado, for example, where MAR occurs in deep confined aquifers, the pressure head at neighboring wells can fluctuate hundreds of feet and affect pump performance.

Rights to divert water for recharge, rights to recharge, and rights to recover recharged water need to be considered during MAR planning and design. This is because in aquifers with uncontrolled extraction, the MAR operator may not be assured of being able to recover sufficient water. Where water trading occurs, MAR can contribute most effectively to restoring over-exploited aquifers.

With respect to recycled water recharge, recharge well sites may need to be located far enough from potable production wells to satisfy regulatory requirements. In California, wells injecting highly treated recycled water need to be located 3 to 6 months groundwater travel time from potable production wells. See Section C.2.2 in Appendix C for more information about California regulations.

At the Orange County Water District Talbert Gap Seawater Intrusion Barrier, which recharges highly treated wastewater (Case Study 8 in Appendix D), no potable production wells can be installed in the vicinity of the barrier to ensure that the minimum travel time requirements are met. Many other states and USEPA also recognize and accept the additional soil aquifer treatment (SAT) provided during storage and conveyance through confined and unconfined aquifer systems, but they do not require a minimum storage period, as California does. Storage underground is considered another public health barrier to protect underground sources of drinking water, effectively providing much longer time for disinfection processes and other water quality benefits.

In the many states where water is not privately owned but is instead held in public trust by the state for reasonable, beneficial use by permittees, a different approach is appropriate for protection of stored water so that it is available for recovery by the permittee when it is needed. Landownership of all property overlying the usually small radius encompassed by the stored water is the surest and simplest approach. Within municipal and county boundaries, a local ordinance may be enacted, providing "institutional control" of the stored water in the selected storage aquifer and within a defined area. Most recharge wells and ASR wells are on small land parcels. A wellhead or well field protection area (WPA) may be negotiated with surrounding landowners or enacted through institutional controls, providing the necessary degree of protection for the stored water.

5.4 PILOT TESTING

The design process often can benefit from conducting a pilot test to evaluate the actual response of the aquifer to recharge after the planned implementation process. The need for a pilot test should be identified in the final stages of the planning process (Chapter 4) or as part of the preliminary design (Section 5.1.2). Results of the pilot test help refine final design elements and identify potential problems and data gaps. Pilot testing should be of sufficient length to allow the facility's response to reach, or at least approach, equilibrium and should incorporate slower processes, such as the leakage of water into an underlying aquifer, clogging, geochemical reactions with slow reaction times, or unidentified subsurface boundaries such as faults. Care should also be taken to factor in the effect of other wells or recharge activities (both natural and artificial) during the test period. Because the results of pilot tests tend to be scale dependent, these tests should be constructed at full scale, or as close to full scale as possible. For example, results of small-scale test projects with small-diameter wells tend to show low flow rates and greater mixing, yielding relatively poorer results than full-scale wells. In the case of ASR wells, pilot testing can be considered "demonstration testing," as it is conducted with a full-size and fully equipped well.

A proposed pilot testing and monitoring program should be developed in some detail. This testing could include monitoring hydrologic parameters, such as streamflows and groundwater levels before the onset of recharge operations, followed by monitoring several wetting and drying cycles for shallow basin recharge or several months of test operations for deep basins or river channel recharge. Testing time should be included to allow for adequate evaluation of potential geochemical reactions revealed by geochemical modeling. For ASR wells, this testing may involve development of the target storage volume followed by one or two cycles of recharge and recovery, whereas for recharge wells, this testing would involve several months of operation under typical conditions. Data collection needs and procedures are included in Chapter 9.

5.5 DESIGN DOCUMENTS

This section addresses preparation of detailed designs and specifications for recharge facilities. This section does not address conveyance, treatment, or other necessary project facilities. Following the types of practices described in this section should increase the probability of project success and help to avoid change orders, delays, and potential construction claims. A typical final design contains the following items:

- Type, size, spacing, and location of surface and subsurface recharge facilities and their appurtenances, including operation and maintenance costs;
- Predicted quantity and quality of the amounts of storage and of project yield;
- Unit and total costs of all facilities, including operation and maintenance costs;
- Schedule of construction;
- Method of meeting required quality by blending, treatment, or use restriction;
- Location, design, and costs of environmental mitigation projects, laws, and regulations affecting the project;
- Water rights, if any, to be acquired; and
- Institutional arrangements to secure benefits and avoid conflict with other water users.

5.5.1 Preparation of Detailed Plans and Specifications. The plans and specifications differ between surface and subsurface MAR facilities in several aspects and so are discussed separately, following.

Surface Facilities. Design plans for surface recharge facilities should illustrate both plan and cross-section views of each key design element. Plans should be as detailed as practical and specify where the contractor has flexibility to use its own means and methods and/or to substitute equipment or materials. Design plans should address and illustrate site grading, roads, basin dimensions, slope stability and erosion control, source water piping, valves, meters, SCADA/communications facilities, weather stations, and all other systems associated with the project.

Construction specifications for surface recharge facilities are typically organized using the Construction Specifications Institute's MasterFormat list of numbers and titles for the construction industry (CSI 2014). Construction specifications for basins, particularly those delivered through a design-bid-build approach, must clearly describe excavation procedures and what constitutes acceptable and suitable fill material, if required. Material used for fill should be tested before placement to ensure that it meets the design specifications, and the specifications should identify which parties are responsible for testing and approval of replacement material. The specifications should also address how the excavated basin bottoms are to be finished to minimize disturbance. If at all possible, the native surface should be used as the recharge surface (referred to as the "cut surface" in Southern California), and low ground-pressure equipment should be used to minimize compaction of the recharge surface. If backfill is required, it should be emplaced by conveyor belts or pushed into place using low ground-pressure equipment. Lastly, the specifications should address how unexpected conditions, if encountered during construction, are to be handled.

Subsurface Facilities. Although subsurface recharge facilities are inherently different from surface facilities, the performance potential of these facilities is equally influenced by site selection, design, and construction methodology. As a result, similar best practices should be followed when preparing detailed plans and specifications. Presumably, site selection and characterization have been performed as part of the feasibility and pilot testing phase (Chapter 4 and Section 5.4), and data collected during earlier phases are available for use in the detailed design phase (Section 9.2).

Subsurface facilities typically incorporate both belowground (well or trench) and aboveground (building, piping, valves, electrical, instrumentation, and control and SCADA) components. These project components are commonly constructed by different types of contractors. As such, separate design plans and specifications should be prepared, and it might be advantageous to bid these packages separately. Alternatively, it may be advantageous to bid them as separate packages in the same bid document, with a general contractor and several subcontractors. The general contractor would then be responsible for ensuring that all of the pieces come together upon completion of construction. Another approach would be for the design engineer to serve as the construction manager, representing the owner's interests and bearing some responsibility for satisfactory project completion.

Belowground components, such as a recharge well, are customarily designed with input from both a hydrogeologist and a civil engineer. Effective integration of engineering and hydrogeology is essential during a MAR project, not only during design but also during construction and testing. If the well is to be used for production and for recharge (i.e., an ASR well), or if a recharge well is to be equipped with a permanent pump to facilitate periodic backflushing, then the structure must be designed to accommodate all necessary downhole components, such as pumping, recharge, and monitoring equipment. Well designs should be illustrated in both plan (surface completion) and cross-section views and should be as specific as possible. For example, the design plans should specify boring diameter; casing material, type, thickness, diameter, length, placement interval, and screen slot size; and type and placement interval of annular fill materials. Special consideration should be given to connection points of dissimilar metals, and selection of casing material should consider the chemical characteristics of both the source water and the native water in the aquifer.

Well pumping and recharge equipment can be included in either the design package for below- or aboveground components. Because the pump is typically installed by a well driller, it is usually best to include facilities for equipping the well with the subsurface components (the pump, motor, column, downhole flow control valve, transducers, and monitoring tubing). Regardless, detailed design of well pumping equipment should not be finalized until the well is installed and fully tested. The choice of equipment is a function of many factors, including the well's production and/or recharge capacity, pumping depth and head, system head and friction losses, characteristics of the power source, and use of a VFD. Pump and motor selection and sourcing must be closely coordinated to avoid problems associated with incompatibility. This issue is particularly pertinent for threaded connections for pumps, column pipe, and downhole flow control valves, because there are many different thread patterns and few uniformity standards. One contractor should be responsible for ensuring compatibility.

Design of aboveground components typically involves a combination of engineering disciplines, including civil, mechanical, electrical, and instrumentation and control, and potentially an architect and structural engineer for building features. In addition, a geotechnical investigation should be conducted if the well facility includes a building or pad. Design plans should illustrate both plan and cross-section views of each key design element and should be as detailed as practical. The plans should illustrate how the site will be accessed, location and nature of electrical power sources, and connection points to existing infrastructure. The site configuration should accommodate access for heavy machinery (including drilling rig and pump rig) and temporary storage areas for pipe and equipment. The design plans should address site preparation and completion features (including grading, fencing, and security); buildings and pads; civil, electrical, mechanical, SCADA, and communications equipment; and how and where the new infrastructure will connect to existing infrastructure.

Similar to the design plans, separate construction specifications should be prepared for below- and aboveground project components. Drilling and well completion specifications must address specialized provisions, such as lost circulation, well plumbness and alignment, casing welds, water and drilling fluids disposal, noise abatement, geophysical logging, well development, and production testing. These specifications should also identify where water can be sourced and fluids disposed of or discharged, where utilities can be accessed and who will pay for them, and how the site must be maintained and left at the time of demobilization.

Construction specifications for aboveground components are typically organized using the CSI MasterFormat (CSI 2014) and should be prepared similarly to those for traditional civil works and water infrastructure projects. The specifications should address procurement and contracting requirements, general requirements, applicable standards, materials testing (including concrete and other materials), equipment, electrical, HVAC systems (if applicable), earthwork, utilities, and anything else relevant to project construction. If the aboveground components are bid separately, and presumably constructed by a different contractor, the specifications should address the interfaces with the belowground components and clearly identify lines of demarcation between the construction packages.

5.5.2 MAR Project Operation and Maintenance Plan. An O&M plan is necessary to evaluate and maintain system performance. Significant effort should be spent on developing an O&M plan, including monitoring and periodic performance assessments. Sections 7.2 and 7.3 provide additional discussion on project O&M for surface and subsurface facilities, respectively.

Performance criteria evaluated and refined during the planning and testing phases should be integrated into a comprehensive O&M plan that is to be diligently implemented and will evolve over the life of a MAR facility. Critical parameters to monitor include source water quality, basin or groundwater levels, and infiltration or recharge rates. Monitoring, reporting, and trigger points need to be established so that critical performance criteria are not exceeded. The importance of establishing critical trigger points cannot be ignored. For instance, if excessive solids enter a MAR well, the well performance can be permanently damaged if the well is not backflushed at the appropriate time.

A recommended approach for ASR wells and recharge wells is to develop a draft O&M plan upon completion of construction of all well and wellhead facilities, using this plan as a training support document for operators. During the first year of operations, performance monitoring provides lessons learned that will be integrated into a final O&M plan to guide subsequent operations. Adjustments to operating procedures, backflushing frequency and procedures, monitoring programs, SCADA set points, alarms, and time delays may all be expected during the first year of operations.

5.5.3 Periodic Review Schedule. At the time a decision to proceed with the MAR project is made, a schedule should be developed for how often after initial construction the design criteria and planned accomplishments of the MAR project are to be compared with the results actually achieved. Consideration should also be given to providing for a periodic review of public attitudes to the MAR project, environmental impacts, technical and economic performance, and the degree to which it contributes to improved water management and supply.

5.5.4 Project Life. A decision should be presented in the final design on terminating the MAR project at a certain time or making the MAR project permanent by proper maintenance and replacement of MAR project features as they wear out. Wells are typically considered to have an economic lifetime of 50 years if they are constructed appropriately. It is appropriate during initial site planning for a recharge or ASR well to consider whether or not to include provision for a future replacement well at the same site.

CHAPTER 6
CONSTRUCTION

The intent of this chapter is to highlight the elements of construction that are most likely to affect the performance of MAR facilities and those elements of construction that are unique to successful operation of a MAR project. It is assumed that at least one member of the project team has construction experience with comparable types of projects, including excavations, earthen dams, well drilling and installation, and conveyance facilities. Therefore, this section does not cover basic concepts related to those aspects of construction. Additional information on construction of surface and subsurface facilities may be found in references provided in the back of this standard.

The primary goal of any MAR project is to provide the most efficient connection between the recharge facility and the target aquifer. The types of connections vary substantially between surface and subsurface facilities; thus, construction-related considerations specific to surface facilities are presented in Section 6.2 and construction-related considerations specific to subsurface facilities are presented in Section 6.3 (Figure 6-1). However, despite the substantial differences in aquifer connections between surface and subsurface facilities, all MAR projects share several basic elements; those are presented in Sections 6.1 and 6.4 through 6.7 (Figure 6-1).

Construction of MAR facilities should not begin until the planning (Chapter 4) and the design (Chapter 5) phases, along with the appropriate data collection and analysis (Chapter 9) have been successfully completed. During the construction phase, the

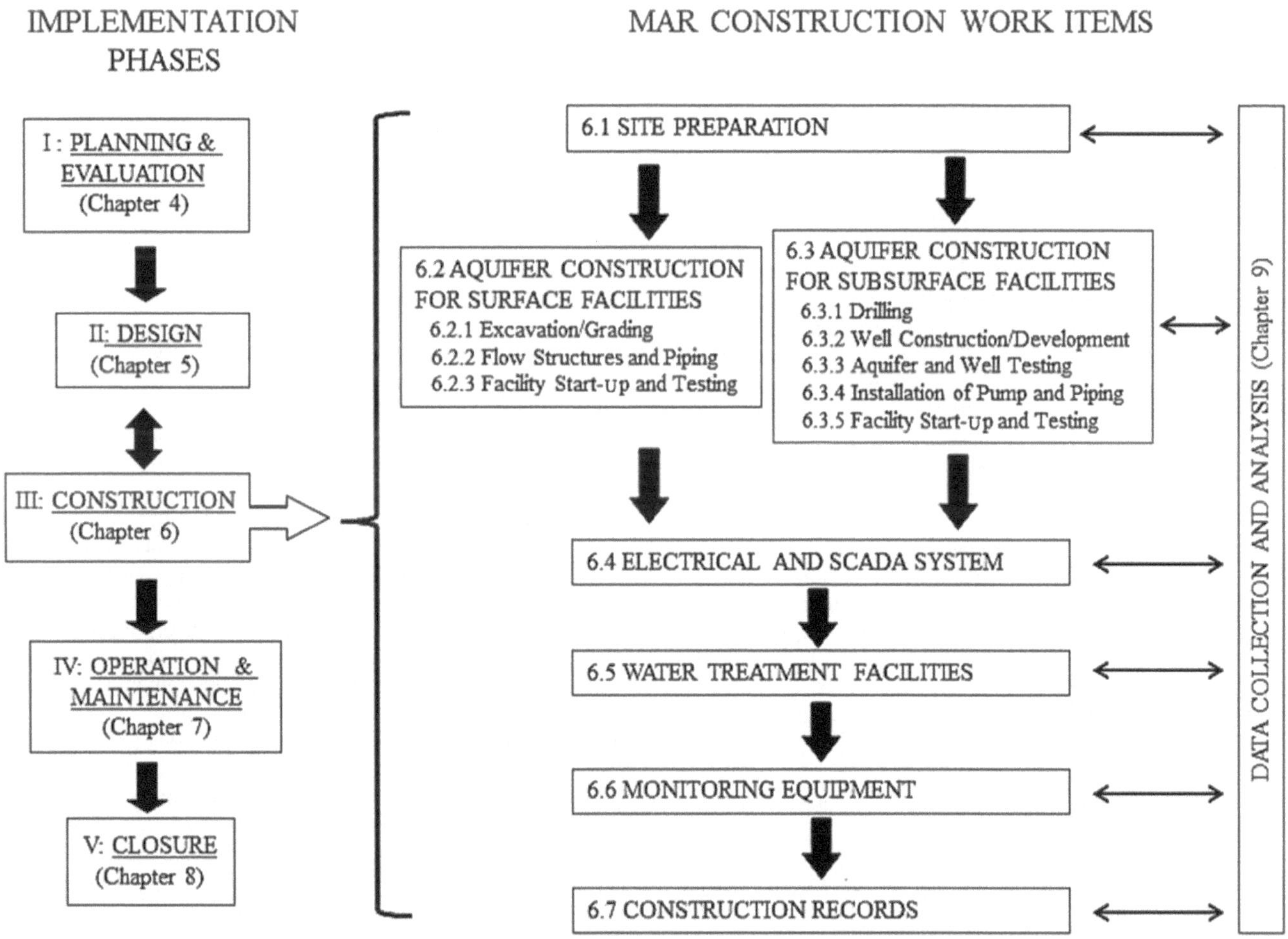

Figure 6-1. Details of the construction phase leading to operation of a MAR project.

project design documents should be followed, and any observed discrepancies should be documented and discussed with the appropriate team members and stakeholders. If necessary, those construction elements should be postponed until the discrepancies are addressed or plans are changed to accommodate changed conditions encountered during construction.

6.1 SITE PREPARATION

Initial site preparation typically involves clearing the site in preparation for excavation, grading, and/or drilling. Site preparation can involve removing or relocating buildings, structures, underground utilities, vegetation, debris, and other items at the site. Surface facilities generally require larger areas than subsurface projects, so these activities may be more extensive for the former.

When working in stream or river channels, site preparation should include routing flows around the work area or making accommodations to work with water present in the channel. Particular care should be taken when site preparation work occurs near wetlands or other environmentally sensitive riparian areas. Federal regulations are in place in the United States regarding dredging or filling wetland areas. It is essential that the MAR project team be aware of and comply with applicable federal, state, and local regulations when undertaking site preparation activities.

Site preparation work should also include a final field inspection to identify any changes that might have occurred since the planning, design, and permitting processes were completed. For example, nesting birds may have relocated, and critical habitat areas may have changed.

As for any construction project, utility clearances must be obtained before any digging or drilling begins. All appropriate permits should be in place and available on the construction site. Permits should include site access, drilling and well or soil boring construction, discharge and disposal permits, stormwater control, and other environmental permits. A complete list of necessary permits should have been developed as part of the design process (Chapter 5). In addition, other work to comply with regulatory and environmental requirements may be completed during the site preparation phase of construction. This work may include the application of various techniques or best management practices (BMPs) to limit the discharges of dirt, dust, and stormwater from the site; to reduce noise; and to establish buffer zones around sensitive habitat.

6.2 AQUIFER CONSTRUCTION FOR SURFACE FACILITIES

Surface MAR facilities that connect source water to the target aquifer include infiltration basins, irrigation ditches, and enhanced infiltration in streams and rivers. Common features of surface MAR facilities are that they are topographic depressions in which recharge is typically governed by gravity flow. Considerations for vadose zone wells (dry wells) are included in Section 6.3 because they are vertical shafts that penetrate the subsurface. The construction techniques required to establish an aquifer connection for surface MAR facilities are relatively straightforward as compared with those for subsurface facilities. The following sections discuss the key elements of construction for a surface MAR project. These elements include site excavation and grading (Section 6.2.1), flow structures and piping (Section 6.2.2), and facility start-up and baseline testing (Section 6.2.3).

6.2.1 Site Excavation and Grading. The main objective of excavation is to remove unwanted sediment, to form a basin or ponding area for recharge to occur, and to expose the target strata while minimizing its disturbance and compaction. The basin should be constructed to allow future access for scraping and cleaning equipment, for example, to provide an access road into the basin for tractors and cleaning equipment.

Where practical, it is recommended that lightweight equipment be used to minimize soil compaction during sediment removal and final grading activities. Heavier equipment can be used to remove materials overlying the target strata, transitioning to lighter weight equipment near the target strata if needed. If compaction occurs during construction, methods such as deep ripping and over-excavation can be used to loosen up the sediment before final grading.

Detailed inspection during excavation is essential to ensure that site conditions are as expected based on data collected during earlier project phases. Changes in the excavation design may be required if subsurface conditions differ from what was expected based on existing data. For example, if unexpected fine-grained layers are encountered within the target strata (including the unsaturated zone beneath a surface facility), then these layers may need to be removed during excavation if practical. Field testing of the target strata, including sieve analyses and infiltration testing, may be required to ensure that the target strata meet design criteria. If the potential for soil contamination is discovered during any phase of construction, detailed inspection and testing may be required to ensure that contaminated soils are removed and properly disposed of.

If excavated sediments are favorable for infiltration, it may be advantageous to stockpile them nearby or on site for later use. These sediments can be used for filling depressions that may form after multiple cleanings of the recharge facility. They can also be used to replace fine-grained layers that are discovered during excavation and that may restrict infiltration. Stockpiled sediment should be of equal or better infiltration quality than the target strata based on grain-size distribution data and, if possible, infiltration data as measured in the laboratory or by other approved methods.

In the event that archeological artifacts or fossils are discovered during excavation, additional inspection may be required. These and any other potential inspection requirements during excavation should have been identified in permits issued for the project.

6.2.2 Flow Structures and Piping. The construction of surface MAR structures, piping, and ancillary facilities should be conducted in a manner that minimizes disturbance and/or compaction of the intended recharge sites. The facilities should be constructed as designed and in accordance with project specifications, and any deviations from the design plans should be documented in the construction as-built drawings.

Energy dissipation structures are often incorporated into surface MAR facilities to improve operations and reduce maintenance. They include structures where water enters the recharge facility and liners on banks that prevent erosion.

Recharge basins should be equipped with access ramps to enable entry of equipment for basin maintenance. Water delivery piping should be constructed with manholes and/or cleanouts to enable access for cleaning and maintenance.

Once construction work is completed and before commencement of recharge activities, it may be beneficial to conduct deep furrowing, disking, or raking (also known as ripping) of the recharge basin to loosen sediments compacted during construction. In addition, piping should be inspected and potentially cleaned to ensure that it is clear of debris or obstructions.

6.2.3 Facility Start-Up and Testing. After surface recharge structures and ancillary facilities are constructed, they should be tested to evaluate and confirm their performance. Facility start-up and testing (including here what is also referred to as baseline testing and commissioning) should include a short-duration test that consists of diverting water from the identified source, transmitting the water through conveyance infrastructure, introducing the water to the recharge structure, and then monitoring the filling and percolation rates. Typical activities during start-up testing include water quality sampling and/or analysis; measurement and monitoring of flow rates, basin water levels, and percolation rates; testing of instrumentation; monitoring of the vadose zone and aquifer near the recharge facility; and observing recharge operations. Any deviations from anticipated performance should be evaluated, diagnosed, and resolved before initiating full-scale operations. A description of baseline performance parameters and relationships and baseline data collection is included in Section 9.9.

6.3 AQUIFER CONSTRUCTION FOR SUBSURFACE FACILITIES

Subsurface MAR facilities that connect source water to the target aquifer include direct injection (recharge) wells, aquifer storage recovery (ASR) wells, and vadose zone wells (dry wells). Common features of subsurface MAR facilities are that they are constructed in vertical shafts drilled into the subsurface with recharge targeting subsurface zones. Vadose zone wells differ from most subsurface recharge wells in that they do not penetrate the saturated zone and typically function by gravity flow. The construction of subsurface MAR facilities has unique requirements that differ from those of surface facilities. This section provides a review of key considerations in subsurface facility construction.

The key elements of construction unique to subsurface MAR facilities include drilling (Section 6.3.1), well construction and development (Section 6.3.2), aquifer and well performance testing (Section 6.3.3), installation of pump and piping (Section 6.3.4), and facility start-up and testing (Section 6.3.5). It is assumed that at least one member of the project team has experience with drilling, well construction and testing, and installation of pumps and piping. Many aspects of well design, construction, and operation are described in guidance documents and textbooks, including Ahmed and Ayoub (2014), ASTM D5786, Sterrett (2007), and Pyne (1995, 2005). Only elements unique to MAR projects are covered in this section.

Throughout the drilling and well construction process, the project team should keep in mind that the goal of a recharge well or dual-purpose (ASR) well is to establish and maintain the most efficient connection with the target zone(s).

6.3.1 Drilling. Design criteria, technical specifications, the contractor selection process, and well development are discussed in Section 5.3.3. Site characterization during the planning phase of the project, described in Chapter 9, provides information needed to plan the drilling program defined in the design phase (Chapter 5). It is, however, common to encounter unexpected conditions when wells are installed, such as shallower- or deeper-than-expected target recharge zones and well yields that are higher or lower than expected. As with all phases of construction, the drilling process for any given MAR project should proceed as clearly defined in the design phase. A robust project design allows some flexibility to accommodate certain unknowns, such as the total depth of the well as it relates to the depth of the targeted aquifer. If unexpected conditions are encountered, the well design may be modified after consultation with the appropriate management team. Any modifications should be clearly documented.

Site characterization continues during the drilling process, which provides the ideal opportunity for collection of additional site-specific information. The detailed information collected during drilling is essential to understanding the hydrogeology of the area. A professional geologist or hydrogeologist should provide drilling oversight and collect cuttings for lithologic description, and mineralogical analysis during the drilling process. Samples should be collected from individual hydrogeologic units for analysis of hydraulic properties and mineral content, at the frequency indicated in the design documents.

Borehole drilling is a critical component of well construction because the borehole provides the most immediate connection to the aquifer. If not done properly, the well may not perform as efficiently and, therefore, as cost-effectively as it could. The drilling method needs careful consideration. Drilling methods are discussed in detail in Section 5.3.3. In general, the drilling method that is selected should take into account the hydrogeologic characteristics of the target aquifer such that formation damage and wall cake are minimized; both can limit hydraulic connection between the aquifer and the well. Drilling muds and additives to drilling fluids should be designed carefully with this issue in mind. The boring needs to be vertically plumb with no severe "dog legs" or curvatures in the borehole, which could later affect pump operations when the pump is installed in the well. Finally, the driller must maintain a stable hole throughout the construction process and avoid creating voids or collapse of the borehole wall that can impede well construction.

Vadose zone wells are drilled and constructed differently than the typical recharge well. They are usually large-diameter [0.9 to 1.2 m (3 to 4 ft)] borings drilled using a bucket auger and completed above the water table at depths ranging from 24 to 46 m (80 to 150 ft).

Upon reaching the target depth for a given well, the selected suite of geophysical logs should be run. A typical basic suite of geophysical logs may include spontaneous potential (SP), electrical resistivity, gamma ray, and caliper surveys. Other logs that may be useful for certain applications include density, nuclear magnetic resonance (NMR), acoustic velocity, and acoustic televiewer (see Section 9.4 for a more detailed discussion of downhole geophysical methods).

Data collected during drilling and geophysical logging should be carefully evaluated in conjunction with the previous understanding of the site's hydrogeology. This information is then used to finalize the well design. If necessary, some elements of the well design (including screened interval, total depth, screen slot size, and gravel pack) may require modification with new and improved knowledge of site-specific conditions.

In cases where water quality is known or suspected to vary substantially by hydrogeologic unit, it may be desirable to perform zonal (packer) testing and water quality sampling of each zone before well completion.

6.3.2 Well Construction and Development. Well construction and testing should proceed according to the design plan, with special attention given to selection and placement of the screened interval (see Ahmed and Ayoub 2014, 1998, Pyne 2005, or Sterrett 2007 for more details). Again, any modifications to the final design should be clearly documented in the construction records.

During well construction, the major activities to be observed are the welding or casing assembly procedure, the proper location of screen sections within the well assembly, and the quantities of materials installed. The artificial filter pack placed adjacent to the well screen interval must be properly sized and installed because it acts as a partial barrier to minimize the entry of sand into the well during pumping and into the formation during recharge. This step is taken because the recharge and recovery process and the resulting flow reversal can mobilize aquifer materials more readily than in a traditional well, in which the flow is always from the formation and into the well screen. Note that if the filter pack is too fine relative to the well screen, sand locking of the well can clog the gravel pack. Conversely, if the filter pack is too coarse relative to the well screen, excess aquifer materials can enter the well and cause the well pump to wear out more quickly. For dual-purpose ASR wells, or recharge wells equipped with a pump for periodic backflushing, a permanent gravel tube should be installed so that additional gravel may be added if the gravel pack settles during operations.

Construction of vadose zone wells involves the installation of screen and blank casing followed by the installation of clean gravel surrounding the casing. Because a typical vadose zone well is much larger in diameter than the casing, there is a large annular space between the casing and the borehole wall. Because vadose zone wells are completed above the water table and cannot be pumped, challenges exist for development of these wells. It is particularly important to site vadose zone wells in alluvial materials that are coarse-grained with little or no fine sand, silt, or clay materials to maximize the recharge rate. One option to facilitate development (and redevelopment, described later) of vadose zone wells is to include injection tubes in the gravel pack outside of the casing. This method allows for recharge of water or other solutions through the gravel pack that can potentially be extracted from the well screen, thereby allowing for the removal of scale and sediment buildup. Even if the injected solutions immediately drain from the gravel pack into the formation and cannot be extracted, improvement of well performance can be achieved.

Because vadose zone well screens are above the water table, recharge capacity can be lost because of air entrainment. This loss can be remedied by incorporating one or more small-diameter recharge pipes consisting of solid casing, referred to as eductor lines, into the well design. The eductor line is placed at the bottom of the vadose zone well screen, allowing recharge water to be introduced slowly into the screen from the bottom. By maintaining a full water column in the eductor line and keeping water within the well screen interval, cascading water can be prevented in the well, and air entrainment problems can be managed.

Once construction is complete, well development is the next step. The goal of development is to remove fine-grained materials from the aquifer formation near the screen and form a gradation of grain size between the formation and gravel pack that allows both recharged and pumped flows to pass easily, improve the specific yield (efficiency) of the well, and permit higher recharge and production rates. The development of the well should be the responsibility of a certified or licensed well driller and is not complete until the specific capacity (the discharge rate or drawdown) stabilizes and both the sand production and turbidity design criteria are achieved. Well development can be conducted using a variety of methods, including mechanical methods such as surging, pumping and bailing, airlift development, and high-pressure jetting with water; sonic jetting; or the addition of chemical additives to disperse drilling mud and clays. In general, it is best to start well development with lower energy methods and then apply progressively more energetic methods. This sequence avoids the potential for damaging the formation or collapsing the screen.

Redevelopment is typically required on a routine basis for ASR wells and may require mechanical or chemical means to remove small particles lodged in the gravel pack and/or aquifer and to remove biofouling and encrustations in the well (see Pyne 2005 for more details). Routine redevelopment should be included in the operations and maintenance plan (Chapter 7), though the method of development and frequency may evolve in an iterative manner based on well performance over time and results from different development procedures. The same equipment and procedures are needed for periodic redevelopment of recharge wells, whether they are equipped with a permanent pump or instead rely on equipment and procedures brought to each well site on a periodic basis to maintain their performance.

Vadose zone wells cannot be redeveloped like wells screened in the saturated zone. For this reason, it is especially important to consider the quality of the water put into vadose zone wells. Advanced treatment that removes suspended particles and microbiological nutrients and disinfects is important to extending the ultimate life of a vadose zone well. Even with aggressive pretreatment, vadose zone wells only have a typical life span of 5 to 10 years. The cost-effectiveness of vadose zone wells should be evaluated relative to that of ASR wells or recharge wells that penetrate to the saturated portion of an aquifer. Cost for the latter is generally higher, but performance and longevity are also greater.

6.3.3 Aquifer and Well Performance Testing. After well construction and development, aquifer and well performance testing should be conducted. Data gathered as part of this testing are used to estimate aquifer properties, evaluate if additional well development is required, and size the pumping equipment. The proper procedures for conducting aquifer tests are described in many textbooks and guidance documents (including ASTM D5786, Kruseman and de Ridder 1992) and generally consist of conducting a step-drawdown test, a constant-rate pumping test, and then a recovery test. For ASR wells and recharge wells, a step rate recharge test is also conducted if an acceptable source of water supply is available for recharge. Preferably some of the step flow rates overlap between the step-drawdown and the step-recharge tests so that ratios of specific injectivity to specific capacity can be calculated. Such data guide design and operation of the equipment provided in the well. The water level data collected during these test phases are used to determine baseline aquifer hydraulic properties, well efficiency, and recharge rates and to observe and quantify groundwater mounding. Water quality sampling is also required to document existing aquifer water chemistry conditions so that these conditions can be monitored during the operation of the facility. Water chemistry is important and is used to troubleshoot any future geochemical, clogging, and facility capacity issues.

It must be emphasized that the initial determination of the well and wellbore hydraulic characteristics is essential to evaluate the well's performance over time. Monitoring programs for recharge and dual-purpose (ASR) wells should be developed to track well performance over time. The initial baseline hydraulic performance of the well is the standard with which all future system performance should be compared.

6.3.4 Installation of Pump and Piping. The well pump and piping should be installed according to the design plans and results of aquifer and well performance testing, with modifications as necessary to accommodate the actual construction details. Particular attention needs to be paid to observation of the correct installation of the following components:

- Injection tubes or pump column recharge where the recharge water level is below the ground surface. These types of facilities can recharge with vacuum conditions in the upper portions of the recharge tube. It is essential for proper performance for these tubes to be able to hold a complete vacuum with zero air leakage because allowing air and oxygen to enter the aquifer in an uncontrolled manner could result in air-binding and promote biological growth and geochemical reactions that could reduce well performance. Testing must be performed on the completed system to confirm vacuum conditions. This approach can work but is "operator-unfriendly" because minor oversights, such as a leaking air-vacuum relief valve, mechanical seal on the pump shaft, or a ball valve on the wellhead assembly that is accidentally left open can quickly lead to air-binding of the aquifer, resolution of which can take days to weeks.
- Downhole flow control valve selection; placement above or below the pump, dependent on site-specific selection criteria, and installation.
- Vertical turbine pump installation, especially where recharge water flows outward through a deeply installed pump. Careful observation of any devices, such as special couplings and antireverse devices, needs to be made.
- Vertical turbine pumps where well alignment issues exist. Setting the turbine pump to minimize bending of the column pipe because of well alignment issues can be critical in some installations. Initial pump testing should confirm that vibration criteria in the technical specifications are met.
- Flowmeter placement to ensure the required upstream and downstream straight pipe distances and to ensure full pipe flow through the meter.
- Wellhead flange should be sealed and hold small positive pressure up to about 200 kPa (30 psi).

6.3.5 Facility Start-Up and Testing. Following facility construction, start-up testing (including here what is also referred to as baseline testing and commissioning) should be performed. Section 9.9 describes baseline performance parameters and baseline data collection. The start-up testing process typically begins by testing each of the facility components or systems individually to evaluate their performance. Any issues must be resolved before proceeding with commissioning activities. The next step is to operate the facility for short periods of time at increasingly higher rates until the design capacity is reached. Attention should be given to any issues with flow reversal in the piping systems, and the resulting potential for particles and turbidity spikes to occur, and any initial well clogging issues. Testing should be conducted using the intended recharge source water and/or recovered water to confirm that the facility is capable of producing water that meets applicable standards over the anticipated range of recharge water qualities.

The start-up and commissioning processes establish a baseline for long-term facility operations. Data and operating experience collected during this process should be captured in a database and in the facility operations and maintenance (O&M) manual(s). The facility can be safely placed into service once commissioning is completed, all issues are resolved, and sufficient data have been collected to confirm that the facility meets design standards.

System testing for ASR wells should involve initial formation and maintenance of a buffer zone, followed by a short-term recharge and recovery cycle, typically of one month or less, to assess that the system components are functioning properly when working together as a system. It is recommended to recharge the well for a month or more, followed by recovery (pumping) of half of the recharge volume, leaving the other half for buffer zone formation. Subsequent operations may include additional short-term cycles if desired, each of which recovers half of the volume stored in that cycle, contributing the other half to buffer zone formation. At such time as the planned buffer zone volume has been achieved, full recovery of the stored water volume in subsequent cycles can be implemented. This is an iterative process, adjusting the target storage volume and the buffer zone volume as needed to ensure that recovered water quality criteria are met. Where local experience or prior geochemical investigations suggest that no geochemical issues are expected, a smaller buffer zone may be sufficient. This zone may be recharged at one time initially, followed by routine operations to store water when available and recover water when needed.

Consideration should be given during the planning process to conducting interim recharge with temporary wellhead facilities during the period (months to a year or more) between completion of well construction and testing and completion of construction of wellhead facilities and equipping of the ASR well. This simple process can often achieve the buffer zone volume before beginning ASR testing, or at least reduce the time and volume needed before achieving ASR performance goals.

One of the most critical outcomes of well start-up testing is determination of the recommended initial frequency and duration of backwash cycles to maintain optimal well performance. This frequency should be closely monitored and adjusted as needed based on changes in well performance and influent water quality. Backflushing is typically required every few days to every few weeks, whether in ASR wells or recharge wells.

6.4 ELECTRICAL AND SCADA SYSTEM

The electrical and automated monitoring or supervisory control and data acquisition (SCADA) systems should be installed according to the design plans, with modifications as necessary to accommodate the actual construction details. Depending on the monitoring system philosophy, it is typical to tune and adjust the control aspects of the SCADA system to operate accurately for the final recharge and pumping characteristics of the well. Particular attention needs to be paid to the correct installation of the following:

- Accurate calibration and verification of instrumentation. In particular, accurate flow and water level measurement and recording. Flow measurements should be to an accuracy of better than ±2% of actual flow, and water levels in the well should be accurate to 30 cm (1 ft).
- Reliable local control and local readouts on flow, pressure, and water levels. This local information is especially required during start-up and any troubleshooting during the life of the well.

6.5 WATER TREATMENT FACILITIES

Water treatment facilities associated with MAR projects consist of both pretreatment and postrecovery facilities and can range in complexity from simple screening or filtration to advanced treatment using membranes and chemical additives. Pretreatment facilities are used for a range of purposes, including filtering, conditioning (such as adjusting the pH or oxygen content of the recharge water, or conditioning the well to control clay dispersion), and/or disinfecting the source water before recharge. The treatment facilities enhance the performance of MAR systems by preventing unnecessary clogging of the aquifer or recharge structure, mitigating undesirable biological or geochemical reactions, and avoiding formation of disinfection by-products (DBPs). Postrecovery treatment facilities are used to ensure that recovered water meets applicable standards before its end use and can include disinfection of the recovered water. The planning, design, and pilot testing, if required, of water treatment facilities must be completed before construction.

6.5.1 Common Water Treatment Issues. Recharge water for MAR facilities should be free from turbidity and suspended solids, and/or treated to the point where it is suitable for recharge without causing frequent or permanent clogging of the aquifer or recharge structure. Low-turbidity water is much more critical for recharge and dual-purpose (ASR) wells, as compared with surface MAR facilities. Particle size distribution analysis of the recharge water can be used to guide pretreatment requirements. Most well recharge operations using sand and gravel aquifer systems have particle sizes below 5 μm. Many operations need to pretreat so that particle sizes are below 1 μm. Drinking water sources for well recharge are by far the most common and typically are in the range of 0.3 to 1.0 μm.

Water for surface MAR facilities is commonly obtained from channeled stormwater or runoff, or excess streamflows to a constructed basin. Source water of a higher quality usually results in reduced facility maintenance and higher overall recharge capacities. Generally, the water quality issue is solids loading and the need to clean recharge basins because of decreased infiltration rates. The cleaning process requires taking the basin out of service, drying, and then cleaning the bottom by scraping, deep ripping, or scarification. Substantial time is usually required for these activities; therefore, the trade-off between higher recharge facility performance and the cost of producing higher water quality needs to be evaluated.

Water obtained for recharge and/or ASR wells usually needs to be of potable quality to not degrade native groundwater quality and to avoid well clogging issues. Clogging can be severe where well backflushing is required daily or even several times a day. Well backflushing removes a portion of the stored water in the aquifer, which can substantially reduce the stored volume and overall recharge capacity if backflushing is conducted frequently and for extended periods.

Another treatment issue is the use of alum or other coagulants to reduce suspended materials in the recharge water. In several cases, the use of alum followed by filtration and then immediate recharge has quickly fouled MAR facilities. Care must be taken to ensure that the treatment process is complete when using coagulants before using the water for recharge. Alum floc post-precipitation and carryover into water distribution systems can occur during high turbidity events in the source water, when alum dosage may be increased. The alum floc then reaches the well and causes clogging, which is difficult to reverse.

6.5.2 Facility Start-Up and Testing. Water treatment facilities also must be fully tested before placing them into service. The start-up testing process typically begins by testing each of the facility components or systems individually to evaluate its performance. The next step is to operate the facility for short periods at increasingly higher rates until the design capacity is reached. Testing should be conducted using the intended recharge source water and/or recovered water to confirm that the facility is capable of producing product water that meets applicable standards over the anticipated range of influent water qualities. The water treatment facility can be safely placed into service once testing is completed, all issues are resolved, and sufficient data have been collected to confirm that the facility meets design standards.

6.6 MONITORING EQUIPMENT

As with all elements of construction, the monitoring plan for both the commissioning period and ultimate operation should have been clearly outlined in the final design phase (Sections 5.5 and 9.7). Details for the monitoring plan should include the number and description of wells or monitoring structures, the types and number of various instruments that are used, the location and placement for each instrument and data logger used, and the timing of installation. The frequency of data collection needs to be considered when purchasing, operating the system monitoring equipment, and storing data. For instance, although data can be collected at a very high frequency, the SCADA system can be programmed to average and store more reasonable and manageable amounts of data.

Monitoring equipment should be installed as described in the project design documents (Chapter 5). If necessitated by changes in the construction plan, the monitoring plan should be modified accordingly; any modifications should be clearly documented. In addition, all monitoring equipment should be installed to allow for easy access during testing and full-scale operations. All instrumentation should be tested, verified, and calibrated once installed.

6.7 CONSTRUCTION RECORDS

Detailed and accurate construction records are important for any facility, and MAR projects are no exception. Baseline testing results and any deviations from the design documents should be included with the construction records. An extensive photographic record, with locations and appropriate details attached to each photo, should also be included in the construction records.

An accurate set of record drawings should be made for the entire facility, with appropriate detail for each component of the facility. For surface facilities, record drawings should include the following:

- A final topographic map of the facility and a stage-storage table that shows the relationship between water level (stage) and volume of storage in the basin. This stage-storage table should be presented at a minimum scale of 3 cm (0.1 ft).
- A table that shows the stage-wetted surface area relationship. This table may require the use of 3D computer-aided drawing (CAD) software. Having stage-wetted surface area data allows for calculations of unit percolation rates [such as meters (feet) per day], which are important when examining facility performance at different elevations and changes in performance over time. Obtaining stage-wetted surface area data is especially important for deeper facilities because the bank surface area represents a larger percentage of the infiltration surface compared with shallow basins.

  For subsurface facilities, record drawings should include the following:
  - Coordination of locations of the well;
  - Elevation of the well and datum used;
  - Lithologic and geophysical logs;
  - Identification of the drilling firm;
  - Type of drilling equipment used;
  - Chronologic record of construction;
  - Gravel pack design;
  - Description and drawing of the well, including all detailed dimensions;

- Alignment and plumbness raw data and analysis;
- Pumping test raw data;
- Water quality sampling and analysis results;
- Disinfection records;
- Construction cost records;
- A downhole video record of the completed well; and
- Start-up test results and performance evaluation.

For all wells, a one-time opportunity exists during the construction phase to collect and assemble vital information on the geology and lithology of the subsurface formations, the materials incorporated into the wells and their configuration, the initial well yield, the aquifer baseline parameters, and the initial quality of the water produced. These data should be assembled in a printed and electronic database and bound in a permanent fashion in a well completion report that documents the complete construction process.

The design drawings, record drawings, topographic site maps, well records, specifications, photographs, and all other relevant information from the construction phase should be organized into the well completion report. Depending on the permitting and regulatory requirements, the completion report may be organized into a single, comprehensive document or it may be subdivided into multiple reports (with appropriate cross-references).

Data contained in a completion report, along with information from the design phase, are essential to creating an accurate record of initial conditions encountered and final construction. The completion report(s), along with other essential information, such as the operations and maintenance plan (Chapter 7) plus ongoing operational and performance data, should be kept in a centralized location so that engineers, operators, maintenance technicians, and others working on the MAR facility can access the records efficiently. To ensure continuity and ownership, it is good practice to assign this responsibility to someone who understands the importance of maintaining these records and how they should be used. These records are essential to assess facility performance over time, especially if the facility does not perform as expected. Much of the information in a report of this nature is helpful and/or required in reporting to regulatory agencies.

CHAPTER 7
OPERATIONS AND MAINTENANCE OF MAR FACILITIES

The intent of this chapter is to highlight the elements of operations and maintenance (O&M) that are most likely to affect the performance of MAR facilities and those elements of O&M that are unique to successful operation of a MAR project. Additional information on O&M of surface and subsurface facilities, including vadose zone wells, may be found in references provided in the back of the book.

The primary goal of O&M for surface and subsurface MAR projects is to provide the most efficient ongoing connection between the recharge facility and the subsurface target zone. The types of connections vary substantially between surface and subsurface facilities; thus O&M issues specific to surface facilities are presented in Section 7.2 and O&M issues specific to subsurface facilities are presented in Section 7.3 (Figure 7-1). However, despite the substantial differences in aquifer connections between surface and subsurface facilities, all MAR projects share several basic elements, which are presented in Sections 7.1 and 7.4.

Regular operation of MAR facilities should not begin until the planning (Chapter 4), the design (Chapter 5), and construction (Chapter 6) phases, along with the appropriate data collection and analyses (Chapter 9), have been successfully completed. A possible exception would be if an aquifer storage and recovery (ASR) well buffer zone is initially formed using interim recharge. Interim recharge is typically conducted using a combination of temporary and permanent facilities during the time between completion of well testing and completion of construction of wellhead facilities, including equipping the ASR well. That period can be several months to a year or more, often depending on lead times for certain well equipment items. During that time, a significant quantity of water can be stored, even though there is no pump in the well to facilitate backflushing to remove any clogging. Initial formation of a buffer zone helps to ensure that water quality criteria are met in the recovered water and also can go a long way toward achieving water supply reliability goals.

In addition, every recharge system should have an O&M manual before start-up. Development of the manual should begin during the design phase, with additions and modifications

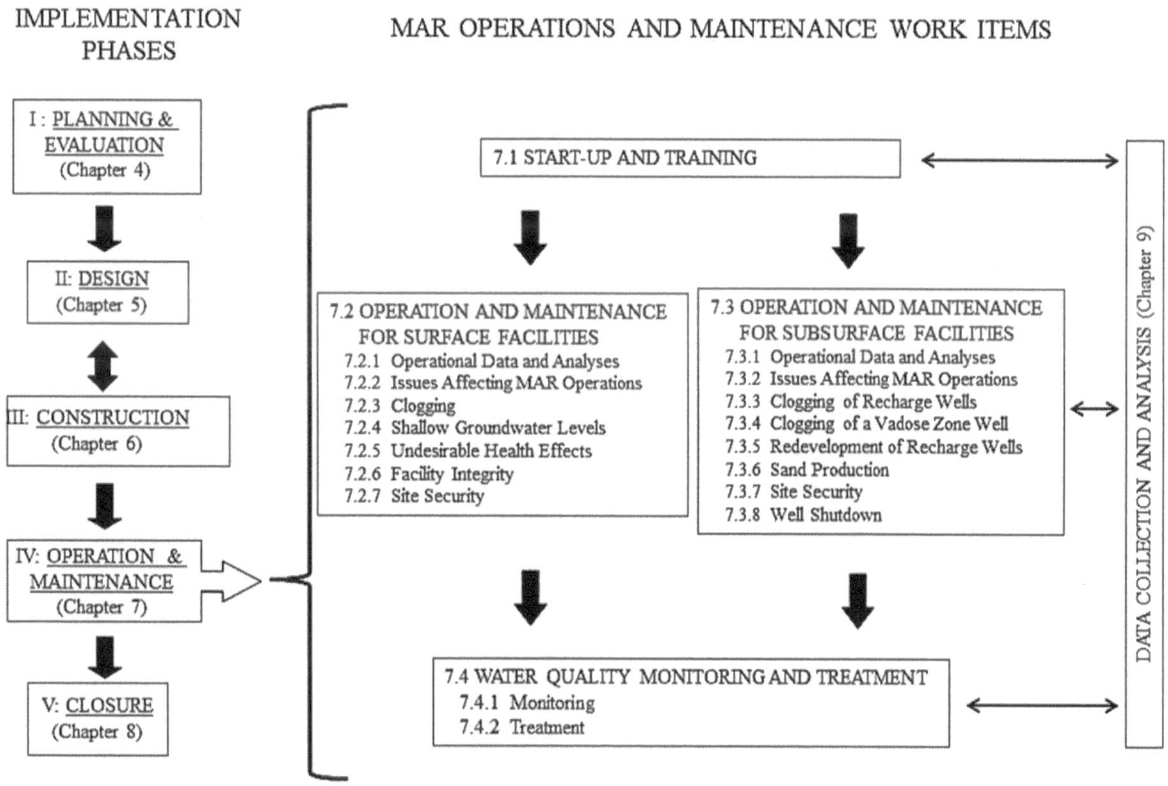

Figure 7-1. Details of the operations and maintenance phase of a MAR project.

taking place as necessary throughout the data collection and construction phases. The manual should include as-built drawings of the facilities, equipment information, equipment and recharge basin maintenance procedures, operations procedures, data collection plan, design of SCADA system, emergency procedures, and a list of key contacts. Before start-up and periodically throughout operation, the manual should be reviewed by the entire project team (including consultants, owner's representatives, and operators); as new issues arise or project objectives change, the manual should be updated as appropriate.

7.1 START-UP AND TRAINING

Start-up procedures should be followed when a new recharge facility is being put into operation, replacement equipment has been installed, or the facility has not been used for an extended period of time. The training of new employees, provision for proper record keeping and procedures, and refresher training for existing employees are essential steps in the start-up process. Key items to consider during start-up of surface and subsurface recharge facilities are discussed in Sections 7.1.1 and 7.1.2, respectively. Operator training is discussed in Section 7.1.3.

7.1.1 Start-Up for Surface Recharge Facilities. Start-up procedures for surface water recharge facilities should include checking the inflow to spreading basins for turbidity or total suspended solids (TSS) to verify that concentrations are below the limits for efficient recharge. Water should be released slowly into surface recharge facilities to prevent erosion and corresponding increases in suspended sediment. It is important that other data, such as inflow rates, water level changes, and visual observations, be recorded so that facility performance can be tracked and evaluated. For instrumented facilities, much of these data are recorded electronically and stored in a computer database. The challenge is linking these data to visual observations and photographs so that they are not lost. For facilities that are not instrumented, the operator should have available facility data sheets and daily operation record forms.

In addition to the recharge facility itself, there are usually ancillary items directly involved with the operation or the recharge unit that should be checked before start-up. These items include the following:

- Diversion structures, such as rubber dams, dikes, and trash racks;
- Valves, such as water supply valves, pressure relief valves, valves on inlet conductor pipes, and flow-diverting valves between spreading basins;
- Flow measurement devices, such as flumes, weirs, and flowmeters; and
- Electrical systems, such as those used for SCADA, control of valves, pumps, measuring devices, computers, security, and lighting.

7.1.2 Start-Up for Subsurface Recharge Facilities. For recharge wells, the facility operator should have an equipment data sheet and a daily well record data sheet for each well to aid in assessing the equipment and aquifer performance during start-up; Figures 7-2 and 7-3 show example forms, respectively. A pump start-up/test data sheet, as shown in Figure 7-4, should be used on start-up. These data can be used to verify the pump curve, calculate the system head curve, set the initial well conditions for future comparisons, and provide valuable data for troubleshooting. Baseline hydraulics should also be recorded during recharge start-up, providing a reference point for subsequent operations so that well clogging and redevelopment procedures can be evaluated and adjusted as necessary.

In addition to the basic recharge well facilities, there are usually ancillary items directly involved with the operation of the recharge unit that should be checked before start-up. These ancillary items are facility specific and vary greatly between recharge wells and spreading basins. Ancillary items include the following:

- Valves, such as water supply valves, pressure relief valves, downhole control valves, air and vacuum relief valves, valves on inlet conductor pipes, and flow-diverting valves between spreading basins;
- Flowmeters, such as on supply lines, between recharge units, on bypass pipes or canals, on backflush discharge pipelines, and on overflow or outflow facilities;
- Chemical feed systems, such as those required for treatment of turbidity, disinfection of recharge water, pH or alkalinity adjustment, or corrosion control;
- Electrical systems, such as those used for control of valves, pumps, measuring devices, computers, and lighting; and
- Downhole transducers and data loggers.

Usually it is advisable to pump (redevelop) a recharge well before introducing recharge water into it if the well has not been used for some time. This pumped water should be pumped to waste until the water quality reaches a stabilized and acceptable level. Similarly, source water for a recharge well should be allowed to waste until quality stabilizes to remove any solids, such as those dislodged from supply lines, before recharge. Any air present in the recharge piping should be vented so that no cascading occurs and so that no entrained air is carried down the well. This method prevents rapid well clogging with entrained air.

During start-up of pumping plants, additional care should be taken with pump shaft lubrication. Open line shafts should be lubricated with water on each start-up if the standing water level in the well is 15 m (50 ft) or more below the surface. Oil-lubricated, enclosed line shafts should receive 5 to 6 drops of oil per minute for each 30 m (100 ft) of pump depth when the pump is initially placed in service. After two weeks of operation, the oil feed rate should be reduced to 3 drops per minute for each 30 m (100 ft) of pump depth (see bulletins of pump manufacturers). Oil-lubricated pumps should be avoided for recharge and ASR wells to minimize adverse water quality effects and the potential for both physical and bacterial clogging.

7.1.3 Operator Training. Operators of MAR facilities need to have a broad skill set, ranging from knowledge of hydrologic principles and facility maintenance to permit compliance and data analysis. Operators of small recharge systems may encompass all the skills required to operate and maintain the facilities, while larger systems may have specialists in several categories. In addition to having the necessary skills, a good operator is genuinely interested in recharge and is constantly learning about how it works, why it is important, and why the recharge system may behave differently at different times.

Case Study 2 in Appendix D describes the current makeup of staff who operate and maintain the Orange County Water District's surface water recharge system and some of the challenges in finding and retaining skilled staff.

In addition to becoming familiar with the O&M manual, additional training on data collection, record keeping, and data analysis is important because keeping good records is essential to any effective and economical groundwater recharge system. Training programs should be developed for both new and

WELL DESIGNATION:

DATE OF INSTALLATION:

PUMP INFORMATION:

SERIAL NUMBER:

MODEL:

NO. OF STAGES:

MANUFACTURER:

PUMP INTAKE SETTING DEPTH:

PUMP RPM:

SUBMERGENCE REQUIREMENT:

LUBRICATION IF VERTICAL TURBINE: OIL/WATER

DIAMETER:

WEIGHT:

LENGTH:

MOTOR INFORMATION:

MOTOR TYPE: SUBMERSIBLE/SURFACE

SERIAL NO.	MODEL:
H.P.	NAMEPLATE AMPERAGE:
MANUFACTURER:	NO. OF PHASES:
FREQUENCY:	RPM:
POWER FACTOR:	
DIAMETER:	LENGTH:
WEIGHT:	

DROP PIPE:

MATERIAL:	JOINT LENGTH:
SIZE:	THREAD:

CABLE (SUBMERSIBLE MOTORS):

SIZE:	VOLTAGE:

WATER LEVEL SENSOR:

AIRLINE MATERIAL	DIAMETER:
DEPTH:	
DOWNHOLE TRANSDUCER TYPE	SETTING DEPTH:

DOWNHOLE CHECK AND/OR DRAIN VALVES:

Figure 7-2. Example well equipment data sheet.

experienced employees. Training for new employees should include why the system was constructed, how the system operates, review of system design, safe work practices, and an overview of the organization that operates the system. Training for experienced employees should include a review of new employee training materials, review of existing procedures and project performance, and reviews of new technical matters related to the operation and maintenance of the particular kind of system being operated. Periodic training of new and more experienced operators is also a good time to review the O&M manual and make needed updates and any other changes required to make it a more useful tool for the operators.

Training on safe work practices should also be included because many tasks are unique to operating and maintaining recharge systems. This training includes working with large volumes of water, both moving and static, earthmoving equipment, high-voltage equipment, large pumps, and dust and sediment within the recharge facilities. In the absence of qualified training staff in house or at nearby agencies, consulting engineers and maintenance or service contractors should be on call for instruction or emergency help.

7.2 OPERATIONS AND MAINTENANCE FOR SURFACE FACILITIES

Operations and maintenance procedures for surface recharge facilities focus on maximizing their long-term performance to infiltrate water at a high rate. Section 7.2.1 identifies operations data collection, data analysis, and evaluation metrics that are helpful in assessing facility performance. Sections 7.2.2 through 7.2.7

MONTH: YEAR: NOTES: WELL LOCATION

(1) DAILY ENTRIES SHOULD BE TAKEN WHILE THE PUMP IS IN OPERATION

(2) ALL WELL AND WATER DEPTH MEASUREMENTS ARE TO BE MADE FROM

DATE	TIME	ON/OFF	Flow (L/s)	TOTALIZER 1000 L	INTERSTAGE (bars)	SYSTEM (bars)	KW	KWH	AMPS	SAND (ppm)	CHEMICAL FEED		WATER LEVEL	WASTE VOLUME		TOTAL (100 L)	REMARKS
														TOTAL	READING		

Figure 7-3. Example of a daily well record data sheet.

discuss some of the most common O&M activities at surface MAR facilities. Water quality monitoring and treatment are discussed in Section 7.4.

7.2.1 Operational Data and Analyses to Evaluate System Performance. Facility operators continuously monitor hydrologic and equipment parameters to understand how the system is performing and to plan for future operations and maintenance. Aquifer water levels and water quality data should be collected before facility operations start to provide a baseline data set against which operational data can be compared.

A trained operator should make daily entries on the daily record forms while the recharge facility is being operated. These values should be checked against the operating parameters established during system design and start-up for the aquifer, the facility, and the facility equipment as entered on the basic data sheets and the start-up/test data sheet.

It is important to monitor all hydrologic and equipment parameters on a daily basis on start-up and for some time afterward. If all systems are operating sufficiently, and the trained operator is comfortable with the operational history of the recharge facility, the time between measurements may be extended. However, changes in quality or rates of flow of the source water may warrant more frequent measurements. Some of the key operational metrics to be tracked include the following:

- Recharge basin and/or facility inflow and outflow,
- Basin water level and storage,
- Aquifer water level and storage trends,
- Projected date when facility recharge basins will be full,
- Projected date when aquifer mounding reaches critical depth,
- Percolation rate and change in rate,
- Projected date when facility will need to be taken offline and cleaned, and
- Projected date when facility will return to operation.

In addition to tracking the day-to-day performance, the long-term performance of the recharge facility should periodically be compared with the predicted performance. Performance trends can be valuable tools in identifying whether current cleaning methods are effective or if remedial actions may be necessary. This assessment can be done by graphing operational parameters versus time to reveal trends in a particular parameter. If the actual performance differs significantly from projected performance, minor changes in operations may be required or changes in predicted recharge rates and quantities should be made based on observed data.

The Rancho California Water District used a multiple-layer numerical groundwater model to simulate aquifer response to historic recharge operations at its Upper Valle de los Caballos Recharge and Recovery Facility. A series of potential operational scenarios were evaluated with the model to increase the overall volume of recoverable recharge water (see Case Study 3 in Appendix D). Sections 9.8 and 9.9 provide additional details on data collection and analysis.

7.2.2 Issues Affecting Surface MAR Operations. Proper maintenance of recharge facilities is essential to maximizing system performance. The ideal approach is to conduct

WELL DESIGNATION:		PUMP:	
TEST DATA:		MOTOR:	
STATIC WATER LEVEL:			(MEASURE BEFORE STARTUP)
LENGTH OF DROP PIPE:		C FACTOR USED:	
SURFACE VOLTAGE:	a/b)	a/c)	b/c)
SURFACE AMPERAGE:	a/b)	a/c)	b/c)
WATER QUALITY RESULTS OF INTEREST:			
TIME AFTER TEST STARTED AND PUMPING RATE WHEN WATER SAMPLE WAS TAKEN:			

TEST VALUES		TEST 1	TEST 2	TEST 3	TEST 4
DATE					
PUMPING RATE (L/s)					
HOURS OF TEST					
TYPE OF TEST					
DRAWDOWN WATER LEVEL					
SURFACE AMPS					
SURFACE VOLTS					
SURFACE (kg/cm^2)					
SAND (ppm)					
SELECTED W.Q. PARAMETER:					
ESTIMATED VALUES					
MINOR LOSSES (m)					
DROP PIPE H. L.					
CALCULATED VALUES					
SPECIFIC CAPACITY					
TDH					
SYSTEM EFFICIENCY					

Figure 7-4. Example of a pump start-up/test data sheet.

preventive maintenance. This approach involves taking periodic action to forestall major repairs or replacement of system components. It is contrasted with operating system components to failure, which results in higher costs in the long run. Preventive maintenance may include drying up and scarifying recharge basins, regular application of lubricants or protective substances (such as oil, grease, and paint to mechanical parts), or replacement of minor parts that are subject to deterioration or repeated failure. Good preventive maintenance requires the operators of the system to be observant and record the behavior of both static and dynamic elements of the recharge system so that changes in their conditions are identified and scheduled for maintenance. Preventive maintenance should be scheduled regularly and may be done when recharge facilities are in operation or shut down, as appropriate.

MAR surface facilities have several potential issues affecting operations that, if not addressed, can cause the MAR facility to not meet operational goals. Key concerns include clogging, shallow groundwater levels, and undesirable health effects. These issues and maintenance activities to reduce their effects are discussed in detail in Sections 7.2.3 through 7.2.5. Other O&M activities are discussed in the remaining sections.

7.2.3 Clogging. The main O&M challenge facing surface recharge facilities is clogging. No matter how free of suspended solids the water source is, every recharge facility becomes clogged and has to be cleaned. Extensive research on clogging at surface facilities has been done over the past 30 years (Hutchinson et al. 2013a, b). Recharge system planners, designers, and operators are encouraged to search the literature and talk with operators of existing facilities to become familiar with clogging mechanisms and to develop an understanding of which mechanisms or combinations of mechanisms have the potential to affect existing or planned facilities. The goal of this section is to briefly introduce some of the more important clogging mechanisms.

Clogging can be caused by physical, biological, and chemical processes. Each of these processes can work individually or collectively to reduce infiltration rates. Many factors influence the development and extent of a clogging layer, including water quality, basin soil texture, ponding depth, hydraulic loading rate, and basin water level.

Physical clogging is caused by the accumulation of suspended solids in the recharge water on the recharge surface. Suspended solids can consist of fine inorganic particles such as silt, fine sand, and clay; algal cells; decaying organic matter; or other organic solids (as, for example, found in wastewater effluent). In-stream recharge facilities that use a constantly moving stream of water tend to result in less clogging compared with terminal recharge basins that are stagnant.

Biological clogging can be caused by microorganisms that grow on the recharge surface and form biofilms that clog the soil by the presence of the organisms themselves and by their metabolic products (such as biopolymers and gases that block pores). Biological growth rates are enhanced by warmer temperatures, sunlight, and greater concentrations of nutrients in the form of dissolved and suspended solids.

Chemical clogging can be caused by precipitation of chemical compounds, such as calcium carbonate ($CaCO_3$), iron (Fe), and manganese (Mn) oxides on and within recharge sediments. Precipitation of these compounds is dependent on chemical and physical-chemical conditions in the recharge water and in the near-surface sediments. For example, water of high hardness can cause precipitation of calcium carbonate in alkaline soils, and high dissolved oxygen concentrations can cause the formation of Fe and Mn oxides in ferruginous soils. It is very important that proper testing of the sediment and potential recharge water sources be checked for chemical compatibility during the investigation phase. Testing may also be necessary if a different water source is to be brought into an existing facility.

The two ways to address clogging as part of O&M activities are (1) minimization and (2) mitigation. Minimization refers to methods that minimize the impact of a particular clogging mechanism. For physical clogging, minimization includes methods to reduce suspended sediment concentrations in the recharge water before it reaches the recharge facility. This step can be done with the construction of settling basins, which allow the larger particulates to settle out before recharge. In the past, it has been suggested that chemical coagulants could be added to the water to reduce suspended sediment concentrations; however, it has been found that residual chemicals remaining in the water after treatment can adversely affect the percolation rate through recharge facility sediments. As a result, it is not recommended that chemical pretreatment be used to reduce suspended sediment concentrations without extensive on-site field testing.

Minimizing the impact of biological clogging can be done by interrupting growth with frequent wetting and drying cycles and by reducing nutrient loading. One way to reduce nutrient loading is to pass the recharge water through constructed wetlands, which are effective in reducing nitrate and phosphorus concentrations. It is also important to maintain an adequate supply of dissolved oxygen in the recharge water to prevent severe anaerobic conditions. Anaerobic conditions, as well as high concentrations of algae, can cause localized conditions favorable to chemical clogging. For example, algae can cause the pH of water to increase because of carbon dioxide (CO_2) uptake, which in turn causes calcium carbonate to precipitate.

It is difficult to minimize the impact of chemical clogging because it involves the chemical makeup of the source water and sediment within the recharge facilities. It has been suggested that if clays within the recharge basin sediments have become dispersed because of adverse reactions with the recharge water such as through cation exchange of calcium (Ca) and sodium (Na), the sediment can be treated with the addition of gypsum or calcium chloride. Detailed pilot testing would be required to ensure that this approach would result in restoring percolation rates and would also be cost effective.

Mitigating clogging helps to restore the capacity of the recharge facility and can be done using a variety of methods, including wet/dry cycling and physical removal of the clogging layer. Wet/dry cycling is commonly used to restore percolation rates for terminal basin facilities and for in-stream facilities that have seasonal flow. The process of drying a recharge facility allows the clogging layer to shrink and crack or decompose, thereby creating breaches in the clogging layer to the underlying, more permeable sediments. The time required for effective drying varies based on the thickness of the clogging layer, the composition of the clogging layer, the permeability of the underlying recharge surface, and weather conditions. After drying, the peak percolation rate is generally lower for each successive wet/dry cycle. The relationship of declining recharge versus different drying times should be determined for guidance in scheduling operations (Detay 1996). This reduction in the percolation rate means that a more aggressive cleaning approach, such as scarification or physical removal of the clogging layer, is eventually required. Sometimes, flooding and drying schedules are determined by other factors, such as breeding of insects, that can be a nuisance to local people or can spread diseases like encephalitis or malaria. Odors from decomposing algae or other organic material can sometimes be a problem. If the basins are also used for recreation, then facility operations may be affected by other concerns, such as the accumulation of floating scum and algae.

Breaking up the clogging layer mechanically by scarifying, ripping, or windrowing is effective; however, these methods do not remove the clogging material from the facility. Over time, the near-surface layer accumulates enough clogging material that it is necessary to remove the layer completely. Physical removal of the clogging layer can be done by hand, using beach cleaners, skid steers, bulldozers, motor graders, scrapers, and other types of equipment. Whatever method is used, it is important to minimize the volume of native sediment removed and minimize the load placed on the recharge surface.

Minimizing the load placed on the recharge surface reduces compaction of finer grained layers and thereby reduces facility performance. Different combinations in size of equipment (load per square meter [square foot]) and number of passes required should be considered to minimize potential compaction. When the cleaning operation is completed, disking or ripping of the recharge surface from approximately 5 to 30 cm (2.5 in. to 1 ft) deep is sometimes used to overcome the compaction that takes place during the cleaning operation. Ripping of the recharge surface must be weighed against the fact that fine material in the recharge water, applied subsequently when operations resume, filters out in the cracks of the deep ripping, making the bottom seal thicker and requiring greater depths of sediment removal to restore recharge rates. Occasionally, a soil sample down to 0.6 to 1.0 m (2 to 3 ft) should be taken to check on the depth of soil compaction that may have occurred during cleaning operations. If compaction is found at these depths, it may be necessary to use a tractor and ripper that can reach these depths. After deep ripping, the recharge surface should be smoothed to reduce the potential problem of deep filtration.

Repeatedly removing the near-surface layer of the recharge facility is not sustainable because it successively removes some of the higher permeability soils and deepens the facility. Therefore, successive removal of the surface layer requires either the periodic importation of new sediment or a washing operation to separate the fine clogging material and coarser native sediments before the coarser native fraction is placed back into the facility. Replacing the removed sediments or washing them can be expensive, so anything that can be done to prolong the useful life of the recharge surface can lead to a more cost-effective operation.

The best schedule of flooding, drying, and cleaning recharge basins to maximize hydraulic loading is determined by local experience, using experimental or pilot systems before the design of the full-scale system, or using the full-scale system itself.

7.2.4 Shallow Groundwater Levels. Recharge operations usually result in an increase in groundwater levels beneath the recharge facility. Mounding of the groundwater surface is caused by water infiltrating faster than it can migrate laterally into adjacent parts of the aquifer. The presence of lower permeability layers located between the base of the infiltration facility and the water table results in shallow groundwater levels much more quickly than if these layers are not present; lower permeability layers should have been identified and their effects mitigated during the planning (Chapter 4) and design (Chapter 5) portions of the MAR project. The groundwater table can rise to as high as about 1 m (3 ft) below the basin bottom without significantly reducing the infiltration rate. When the depth to groundwater begins to be less than 1 m (3 ft) below the basin bottom, the capillary fringe may start to reach the basin bottom and infiltration rates will decrease significantly as the water table rises.

The best way to manage groundwater mounding and to minimize its adverse effects is to adjust the rate at which water is supplied to the recharge basins. The rate at which a groundwater mound rises and its lateral extent are dependent on the volume and rate of applied water, the configuration of the MAR infiltration facilities, and the subsurface geology, including the lateral hydraulic conductivity of the aquifer, the depth, extent, and vertical conductivity of the confining layer, if present, and the height and slope of the recharge mound.

O&M activities should include the installation of shallow wells or piezometers near the infiltration facilities, the routine monitoring of water levels, and evaluation of those data to characterize the mounding. Groundwater mounding below infiltration facilities also can be predicted mathematically, using analytical methods (see Glover 1960, Hantush 1967, Bouwer 1978) or by computer modeling using one of many groundwater models (see Chapter 9).

7.2.5 Undesirable Health Effects. Excessively long flooding periods or poor maintenance practices (such as insufficient cleaning or growth of algae or weeds) may result in production of odors or the development of undesirable insect communities, including mosquitoes. Disease can be spread by mosquitoes from poorly maintained ponding areas. The effectiveness of surface infiltration and soil aquifer treatment to remove viruses and other pathogens needs to be monitored by a well-designed water quality sampling and analysis program to minimize public health problems.

Another potential issue is the unintended release of materials sorbed onto subsurface soils if there is a significant change in the source water used for recharge operations. An example of this was studied for the Montebello Forebay Spreading Grounds in Pico Rivera, California. These spreading basins have used tertiary treated recycled water for recharge since 1962. The Water Replenishment District of Southern California (WRD), the facility operator, considered using a higher quality pretreated recycled water as its recharge water. Studies indicated that using 100% advance-treated recycled water without blending with tertiary treated water had the potential to liberate the bound total organic carbon (TOC) into the shallow groundwater that had accumulated from previous recharge operations. To address that issue, the WRD plans to blend the tertiary and advance-treated recycled water before recharge (see Case Study 6 in Appendix D).

7.2.6 Facility Integrity. Any time water is placed into the recharge system, whether it is impounding the water behind a dam or dike, running the water through channels and pipes, or filling a recharge basin, the entire area receiving water should be checked for leaks. Some leakage is permissible if the water does not carry any soil material. Loss of soil material could lead to erosion, undermining, or possibly a failure.

If leakage occurs around an abutment, or through or under an embankment, and it is carrying soil particles, then prompt corrective action needs to be taken. The first action should be to slowly lower the water level in the recharge facility until the leak almost stops and then to carefully look for the leak entrance on the upstream side of the abutment or embankment. The water level should be lowered slowly because rapid lowering of the water level could cause a failure of saturated sediments upstream of the leak. The leaks must always be plugged on the upstream side of the leaking abutment or embankment. Attempts to stop the leakage on the downstream side are usually unsuccessful because the water finds a new path around the repair or causes a buildup of pressure and makes failure more certain.

7.2.7 Site Security. Recharge facilities can be attractive to members of the public. O&M activities should include appropriate security procedures to protect the public from the potential hazards posed by recharge facilities, such as drowning of both humans and pets. The security procedures should also protect the facilities from vandalism. Examples of security procedures include routinely inspecting facility premises; verifying that fences, buildings, and other structures that should be locked are so; and ensuring that surveillance equipment and security alarms are in proper working order.

7.3 OPERATIONS AND MAINTENANCE FOR SUBSURFACE FACILITIES

Operation and maintenance procedures for most subsurface recharge facilities are centered on maximizing their long-term performance to inject water at a high rate; they are the focus of this section. Section 7.3.1 identifies operations data collection, data analysis, and evaluation metrics that are helpful in assessing facility performance. Sections 7.3.2 through 7.3.8 discuss some of the most common O&M activities at subsurface MAR facilities. Water quality monitoring and treatment are discussed in Section 7.4.

7.3.1 Operational Data and Analyses to Evaluate System Performance. Facility operators continuously monitor hydrologic and equipment parameters to understand how the system is performing and to plan for future operations and maintenance. Aquifer water levels and water quality data should be collected before facility operations start to provide a baseline data set with which operational data can be compared. A trained operator should make daily entries on the daily record forms while the recharge well facility is being operated. These values should be checked against the operating parameters established during system design and start-up for the aquifer, the facility, and the facility equipment as entered on the basic data sheets and the start-up/test data sheet.

It is important to monitor all hydrologic and equipment parameters on a daily basis on start-up and for some time afterward. If all systems are operating sufficiently and the trained operator is satisfied with the operational history of the recharge facility, the time between measurements may be extended. However, changes in quality or rates of flow of the source water and recovered water may warrant more frequent measurements. Some of the key operational metrics to be tracked include the following:

- Recharge system inflow and outflow,
- Recharge well and/or facility inflow and outflow,
- Aquifer water level and aquifer storage trends,
- Projected date when water level in an individual well reaches its maximum,
- Projected date when aquifer mounding reaches critical depth,

- Recharge rate and change in rate,
- Projected date when facility will need to be taken offline and rehabilitated, and
- Projected date when facility will return to operation.

In addition to tracking day-to-day performance, the long-term performance of the recharge facility should periodically be compared with the predicted performance. Performance trends can be valuable tools in identifying whether current cleaning methods are effective or if remedial actions may be necessary. This identification can be done by graphing operational parameters versus time to reveal trends in a particular parameter. If the actual performance differs significantly from projected performance, minor changes in operations may be required or changes in predicted recharge rates and quantities should be made based on observed data. Additional details on data collection and analysis are provided in Sections 9.8 and 9.9.

For ASR wells, cumulative storage volume is monitored. A suggested approach is to delete from the monitoring records the volume in the buffer zone so that operators are less inclined to continue pumping when the stored water volume has been fully recovered. Experience has shown that recovery of water in the buffer zone can often yield water quality that exceeds drinking water standards for some constituents. It is inappropriate to pump out the "walls of the tank" of the buffer zone just because the tank (the injected water) is empty.

7.3.2 Issues Affecting Subsurface MAR Operations. Many issues can affect operation of subsurface MAR facilities that, if not addressed, can cause the facility to not meet its goals. The level of concern for individual issues is site specific and is based on how likely it is that an issue could reduce the ability of a typical subsurface MAR project to reach its goals.

Clogging, backflushing frequency, compatibility of the injected and native groundwater, source water that can vary in quantity and quality, and water quality requirements are perhaps the most common problems encountered during subsurface MAR operations. These issues and maintenance activities to reduce their effects are discussed in detail in Sections 7.3.3 through 7.3.5. Other O&M activities are discussed in the remaining sections.

An example of a project that experienced loss of performance because of clogging is the Denver Basin Aquifer Recharge Demonstration Program included as Case Study 7 in Appendix D. In this case, a well originally drilled in 1988 as a water supply well in the Denver Basin was converted to an ASR well. From 1992 until 1996, 34 recharge and recovery cycles were performed. The specific capacity of the well gradually decreased from the original 8.8 L/min/m (7.6 gpm/ft) to less than 3.5 L/min/m (3 gpm/ft). It was eventually determined that the loss of specific capacity was the result of iron-related bacteria (IRB) that thrived in the well because of the mixing of aerobic and anaerobic water. The well design had included 20 separate screened intervals separated by blank casing. Because the gravel pack behind the blank portions of casing could not be rehabilitated, there was no way to significantly improve the well's performance. The possibility of unwanted biological growth in gravel pack behind long sections of blank casing should be taken into consideration in well design.

7.3.3 Clogging of Recharge Wells. Recharge wells in alluvial aquifers may become clogged quite easily as the water–soil interface outside the well is essentially a cylindrical sand filter. The velocity of flow and the volume of water passing through the filter are comparable to the lower range of flow rates for conventional rapid sand filters. Clogging is believed to occur because of biological growth, filtration of suspended solids, chemical reaction, gas binding, and perhaps other reasons. Wells in limestone and volcanic formations are not subject to clogging to the same extent as those in unconsolidated aquifers because of the reduced level of filter action.

Minerals can precipitate on the well screen, in the gravel envelope (filter), and in the aquifer adjacent to the recharge well because of the interaction of the recharge water and the native groundwater or the aquifer material. The major problem with recharge wells is clogging of the aquifer around the well, especially at the circumference of the borehole (interface between gravel envelope and aquifer), where suspended solids can accumulate and bacterial growth tends to be concentrated. Recharge wells are much more vulnerable to clogging than surface infiltration systems because higher infiltration rates into the aquifer around the borehole per unit area can result in the accumulation of more suspended solids than in surface infiltration basins. In addition, remediation of clogging in recharge wells is much more difficult than in surface infiltration systems, which can be dried and scraped mechanically. Use of dual-purpose ASR wells that recharge at relatively low pressures and are backflushed (pumped) regularly usually reduces or eliminates these problems (Pyne 2005).

If the recharge water contains chlorine but also nutrients and degradable organic carbon, biological activity and clogging may develop in the aquifer some distance away from the well where the chlorine levels have dissipated. This clogging away from the well is difficult to control but may not be significant if it does not affect well recharge rates. Clogging parameters, such as the assimilable organic carbon content of the recharge water and clogging in test columns with much higher velocities than in the actual recharge well system, are useful for identifying relative clogging potentials of various waters. Declines in recharge rates for actual wells are often erratic, seasonal, and sensitive to small changes in water characteristics. A suggested approach is to include in the wellhead facilities design a small bypass filter that can be visually inspected on a regular basis. If it stays white, recharge should continue. If it suddenly changes color, a slug of particulate material has just gone down the well, contributing to well clogging and probably necessitating a backflush. Experience suggests that well clogging events tend to be intermittent rather than continuous. Flow reversals in distribution systems, such as those that occur because of routine line flushing and major fires, can stir up settled sediments in the pipelines.

Reclaimed wastewater used for recharge requires special attention. Recharge tests using reclaimed municipal wastewater at Riverhead, New York; Nassau County, New York; Scottsdale, Arizona; Orange County, California; Hyperion (Los Angeles), California; and Palo Alto, California, have indicated that removal of all organic matter provides maximum protection to the recharge wells from the standpoint of clogging in sandy aquifers. Tertiary treatment is required that may involve application of the latest technology for advanced water treatment, such as activated carbon beds or reverse osmosis. El Paso, Texas, used reclaimed water with tertiary treatment to recharge the Hueco Bolson aquifer, a transboundary aquifer shared between the United States and Mexico since 1985 (Knorr and Cliett 1985, Sheng 2005). This project used recharge wells first and later added infiltration basins after 2001. Currently, they operate both recharge wells and infiltration basins. More information can be found in Case Study 11 in Appendix D.

Suspended Material. Because of the nature of the filtering action of alluvial aquifers, water used for recharge ideally should be completely free of sediment. Because recharge water is almost never free of sediment, periodic redevelopment of recharge wells and periodic pumping of ASR wells are required to remove deposited sediment from a well. A suggested approach is to equip recharge wells with pumps. Suspended solids carried in the injected water may cause clogging at the interface between the filter pack and the well casing, in the filter pack around the well, at the interface between the filter pack and the aquifer formation, and within the aquifer itself. Depending on the particle size, formation material gradation, and flow velocities, sediment may be filtered out immediately adjacent to the well face or may be carried into the formation. If not removed during development of the well after drilling, fine-grained particulate matter may be eroded from the residual layers of drilling fluid on the sides of the drilled borehole or native particles from within the formation itself. These particles are effective clogging material. Also, there may be reorientation of the formation particles into a denser, less permeable pattern if the recharge water velocities are too high. Another potential source of fine particles is dissolution of minerals in the aquifer, such as feldspars, because of incompatible recharge water quality. Conditioning of the recharge water quality may be needed to resolve this contribution to well clogging.

Case Study 8 in Appendix D, Orange County Water District's (OCWD's) Talbert Gap Seawater Intrusion Barrier, describes how, even though a very pure, high-quality source water is used for injection, clogging because of particulates, mainly from the conveyance pipelines, requires routine backwashing and redevelopment to maintain performance (Figure 7-5).

Chemical Reaction. Clogging of recharge wells because of chemical reaction may occur at the screen or casing perforations, the well–formation interface, or in the aquifer itself. Chemical clogging can be caused by (a) precipitated metabolic products of bacteria including iron hydroxide, ferrous bicarbonate, metallic sulfide (sulfur), or calcium carbonate; (b) chemical interaction of the dissolved chemicals in the injected water and in the aquifer formation yielding precipitates, or solution and redeposition of soluble compounds such as gypsum; and (c) reaction of high-sodium water with soil particles, causing deflocculation and swelling of the soil (clay) particles. Well clogging has also been noted because of calcium carbonate precipitation around a pump that barely fit inside a well casing. The head loss as water flowed through the very small annular space, and the calcium carbonate in the recharge water, caused a rapid loss in well recharge and recovery rates.

It has been found that water prepared for domestic service, which has a stabilized calcium carbonate content, may destabilize and may cause some precipitation as the water moves through the gravel pack and aquifer immediately adjacent to the well casing. The deposition of calcium carbonate appears to occur because of the large surface area provided by the gravel pack and the material in the aquifer. At the Los Angeles County, California, Talbert Gap seawater intrusion barrier project (Appendix D, Case Study 8) some success in resolving this problem has been obtained by lowering the pH with acid injection. Adjustment of pH has been used at some ASR well sites to reduce the potential for iron and manganese precipitation.

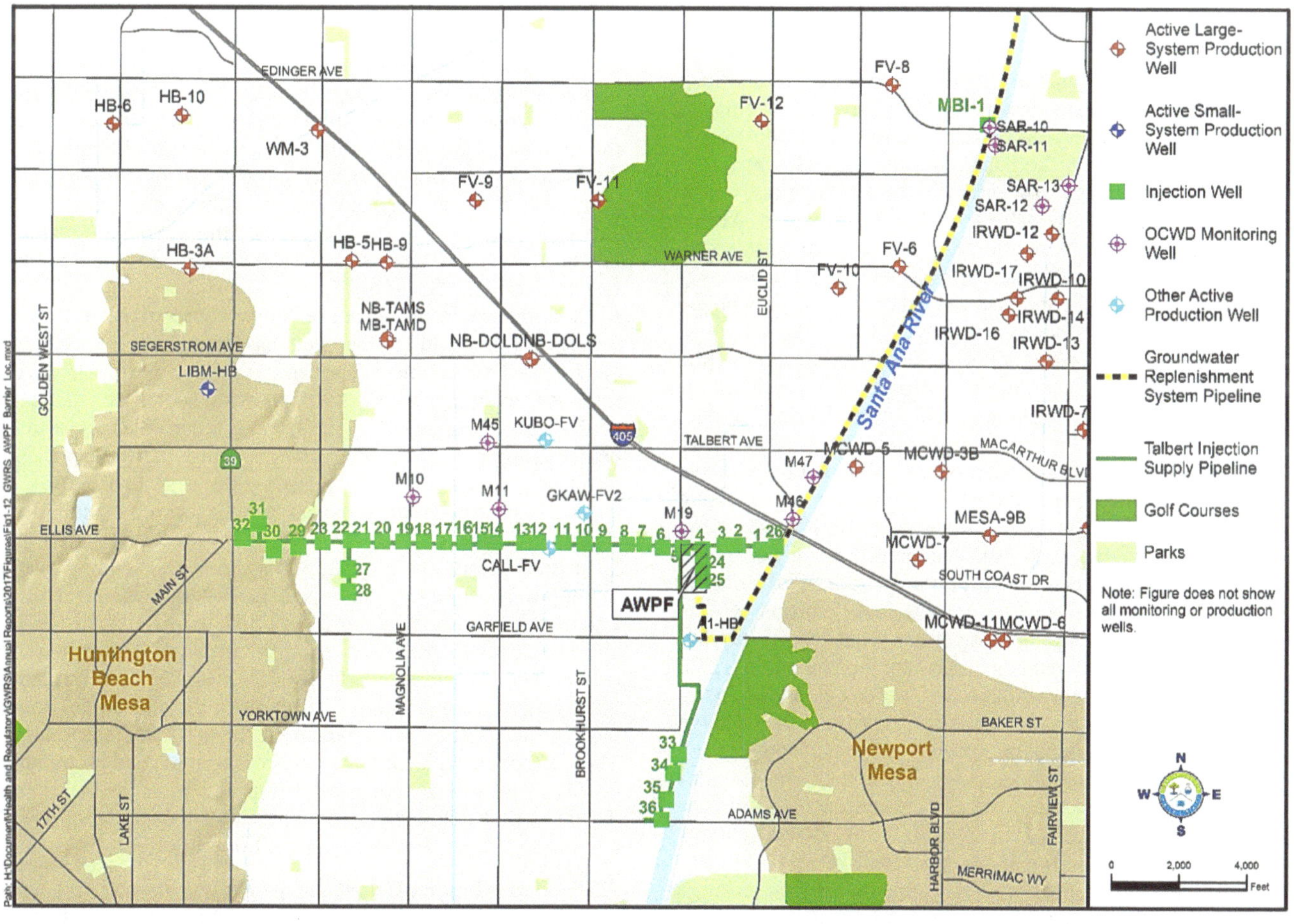

Figure 7-5. Talbert Gap Seawater Barrier location map.
Source: Orange County Water District (2018).

In some alluvial aquifers, expansion of clay materials may occur upon contact with water that has different chemical characteristics than the native water. This expansion typically happens when formation materials containing base-exchange clays are flooded with waters containing high amounts of sodium. The clay material deflocculates, decreasing the size of the interstitial openings, and effectively clogs the formation.

Air-Binding. Admission of free air as bubbles in the injected water forces the bubbles into the interstices of the pore spaces of the aquifer, reducing its hydraulic conductivity and effectively clogging the formation. Such clogging may result from gas coming out of the solution from the injected water when the temperature of the injected water is less than the temperature of the native water in the aquifer, and when the injected water is supersaturated with respect to dissolved gasses, such as because of cascading of water down the well casing or the pump column. The resulting bubbles block the pores of the aquifer, reducing the hydraulic conductivity and, hence, the recharge rates. For this reason, dissolved and entrained air concentrations in the recharge water should always be small. Free-falling water in the recharge well should be avoided to prevent air entrainment. If recharge occurs under a vacuum at the wellhead, as is sometimes the case in ASR wells, care is required during design, construction, and operation to ensure that no leaks occur that could entrain air. If air-binding occurs, extended pumping of the well may be required to restore yield, in addition to providing time for air to dissolve into the groundwater. This process may take days to weeks. Therefore, designing and operating the well to prevent air entrainment is quite important.

Biological Growth. It may also be desirable to provide an initial disinfection of the aquifer formation around the recharge well by injecting a small volume of highly chlorinated water. Such disinfection may prevent growth in the aquifer of any native bacteria that would otherwise be activated by the introduction of an oxygen source and a food supply in the injected water.

As a case study example, in early tests in Los Angeles County, California, a chlorination rate of 8 to 12 mg/L was used. After a period of testing, the chlorination rate was reduced in stages to 1.5 mg/L. Supplemental chlorination was terminated in the early 1970s. Operations have continued since then, relying only on the 0.5 mg/L chlorine residual in the delivered recharge water. At this location, it appears that, after an initial disinfection of the formation in the vicinity of the recharge wells, control of bacterial growth can be maintained at a lower chlorination rate.

The use of chlorine for disinfection should be evaluated with respect to whether it causes adverse reactions with the native groundwater. The biggest concern is production of excessive trihalomethanes (THMs) and haloacetic acids (HAAs) in the groundwater, although such concentrations tend to decline or disappear during aquifer storage, primarily because of bacterial activity and possibly because of other mechanisms. HAAs attenuate within a few days because of aerobic microbial activity, usually resulting from recharge of water containing oxygen. THMs typically attenuate within a few weeks because of anaerobic microbial activity that develops after oxygen in the recharge water has been consumed by aerobic microbial activity and geochemical reactions around the well. Where the storage aquifer contains water with baseline oxygen concentrations, THM attenuation does not occur or is delayed until downhole conditions become more favorable for anoxic microbial activity to occur. This can occur because of well clogging that drives recharge water into deeper storage intervals that may be anoxic or because of oxygen consumption in shallower storage intervals that cause the aquifer to become anoxic. It is possible that high concentrations of chlorine introduced with recharged water could corrode steel casing material in the well, which could cause clogging. Adequate purging of chlorinated water in a well is necessary to avoid generation of THMs, HAAs, and corrosion of steel casing.

When the water supply used for recharge is being treated with chloramine rather than chlorine, bacteriological clogging of recharge wells has generally increased. This increase is probably the result of two factors: (1) the chloramines are not as effective as the free chlorine, and (2) the excess nitrogen in the water probably results in nitrification and is a nutrient source for bacteria.

During periods when no recharge or recovery is occurring, it is advisable to continue a trickle flow of chlorinated water into the well, typically 7 to 37 L/min (2 to 10 gal./min) to maintain disinfection of the well. The trickle should be at least sufficient to replace the well volume in one day, which is about the time a chlorine residual takes to dissipate. Wells that maintain a trickle flow during extended storage periods tend to have few clogging problems requiring extensive well redevelopment. Normal periodic backflushing is sufficient. Some recharge or wells that do not maintain a trickle flow do not have such problems. Most wells that have problems requiring extensive redevelopment procedures do not maintain a trickle flow during long storage periods.

7.3.4 Clogging of a Vadose Zone Well. The main problem with using vadose zone (dry) wells for groundwater recharge is that the wall of the well can become clogged. When clogged, it is practically impossible to remove the clogging material from the inner surface of the well and construction of a replacement well becomes necessary. As the well is in the unsaturated zone, there is no groundwater flow into it, removing material from its inner surface. Clogging may be prevented or minimized by not allowing turbid water to flow into the well. Another clogging problem arises from the slaking and sloughing of clay layers through which the well was drilled; this adds to the turbidity of the water in the well, which clogs the wall of the well in the permeable zones as the water infiltrates the surrounding aquifer. The problem can be reduced by protecting or shielding the clay layers from the infiltrating water through proper well construction (see Section 6.3.2).

Clogging can occur in vadose zone wells from materials in the borehole walls and in the well casing falling into the well (slaking). This problem can be avoided by one of several methods, including lining the well with geotextile, filling the well with sand or fine gravel, or using a perforated pipe in the center to apply the water for recharge. Where there is a pocket of fine-textured layers above a deeper zone of sand or gravel, the sides of the fine-textured layers can be covered with plastic sheeting. The water used for recharge in vadose zone wells must be treated before being applied to remove all clogging agents, including suspended solids, organic carbon, nutrients, and microorganisms. It also must be disinfected to maintain a residual chlorine level to minimize bacterial growth. If clogging still occurs (and long-term clogging is always a possibility), it will then mostly occur because of bacterial cells and organic metabolic products such as polymers on the well wall (biofouling). Thus, although such clogging cannot be remediated by pumping or cleaning or redevelopment, it may be possible that a long drying period could produce significant biodegradation of the clogging material to restore the well for recharge.

7.3.5 Redevelopment of Recharge Wells. For recharge wells, clogging effects are remediated by periodic pumping of the wells to reverse the flow and dislodge clogging materials that are then pumped out. When recharge wells are pumped, the first water extracted often is brown and odorous, and it must be discharged, treated as wastewater, or recycled through the water treatment plant. This situation is particularly true for mild steel cased wells. Pumping schedules may vary from a few minutes each day to a few times per year, depending on how fast recharge rates decline. If pumping does not restore recharge rates, more intensive redevelopment of the well is necessary.

For single-purpose recharge wells, redevelopment methods include the following:

- Bailing and surging;
- Pumping and surging with an air-lift pump;
- Pumping and surging with a deep-well turbine pump;
- Use of dual packers to isolate sections of perforations combined with air-lift pumping and surging;
- Introduction of an acid, then surging and removal of the acid;
- Mechanical brushing of the well screen; and
- High-velocity jetting with water.

Redevelopment of a recharge well by pumping with a deep-well turbine pump not already in the well involves expense and difficulty, primarily because of water disposal problems and the amount of equipment involved. A suggested approach is to equip recharge wells with pumps. The design and equipping of a recharge well would then be the same as for an ASR well. Only the operation of the respective wells would be different. For recharge wells, the cumulative storage volume would steadily increase, with recovery eventually at other wells. For ASR wells, the pump is already present, and the cumulative storage volume will fluctuate between zero and the target storage volume, depending on whether the well is being recharged or recovered to meet seasonal, drought, or other water needs.

The redevelopment procedure used at each well is dependent upon the method used to construct the well and its operating experience, such as known tendencies to yield fine sands and silts and historical evidence of caving around the well casing. In the case of gravel-packed wells, additional gravel may have to be added to the top of the gravel column through a gravel conductor pipe as redevelopment proceeds so that the formation of voids underground is minimized. If a conduit to the top of the gravel pack was not provided in the original design, the possibility of adding one should be investigated. Redevelopment of a recharge well is similar to the development of a new water well and has the same objective (Fowler 1996).

If well development techniques do not solve the problem, the pump should be removed and downhole video (or film) and geophysical logging should be used to determine the source of the problem.

As a case study example, gravel-packed, reverse rotary drilled, asbestos cement pipe-cased recharge wells in a Los Angeles County, California, project have been operated for an average of 20 years. The frequency of redevelopment for these wells varies from yearly to as long as 15 years, with the largest number of wells having a frequency of 2 to 5 years. However, the system has experienced significant loss in capacity at this long frequency. In another example, the Goleta Water District in Santa Barbara, California, has used production wells for recharge of the groundwater basin since 1983. No special redevelopment has been necessary as the production of water from the wells has returned the recharge rate to nearly the original value after each recharge season. There appears to be a slight reduction, however, indicating that redevelopment may be required at some future date. At more than 200 other ASR sites in the United States, long-term recharge decline has not been observed. Back-pumping frequencies at these ASR sites typically range from every few days to every few months.

7.3.6 Sand Production. Sand production is the presence of sand in deleterious amounts in the water produced from a well. Any sand production from water wells is undesirable, and the amount of sand that is considered nondeleterious depends on the use of the water. Certainly, an amount of sand that causes rapid pump wear would be unacceptable. Sand production may create underground voids in an aquifer. If these voids grow large enough, a collapse could occur, damaging the well casing or the inlet section and causing local land surface subsidence around the well. Proper design, construction, and development of a well should result in sand-free water.

If significant amounts of sand appear in the water pumped from an ASR or production well, possible causes that should be investigated are

- Casing may be failing;
- An originally open borehole well may have started to produce sand, perhaps because of dissolution of sandstone;
- Aquifer may be over pumped and velocity of flow in aquifer is high; and
- Gravel pack may be inadequate, insufficiently developed, or may need augmenting.

A sand-free well is dependent on the existence of a properly designed and efficient filter being built up in the annulus surrounding the well inlet. This filter may be an artificial gravel pack or a filter of natural gravels or sands. A well in consolidated formations, such as limestone and sandstone, is often drilled and completed as an open hole with no casing or screen. When these wells produce large amounts of sand, they are often abandoned, recased, screened, or redrilled in another location. Measuring depth to the top of the gravel pack periodically can signal when the gravel pack needs augmenting to prevent short-circuiting of formation sands into the well. Measuring and recording sand production on start-up and after each time a water supply well or ASR well is pumped provides a warning that pump wear may become a problem and sand production is increasing.

Methods of reducing sand production in ASR wells range from reducing the rate of pumping to the abandonment and replacement of the well. The symptoms can often be alleviated by installing sand separation devices on the discharge of the well or downhole on the well pump suction, thus prolonging the life of the well when the underlying problem must be addressed. When failure of the well screen or perforated casing is the suspected cause of sand production, the well should be inspected by a downhole camera, if possible. If the failures are localized, it may be feasible to seal off the failed areas with a liner or by grouting. In severe cases, a smaller diameter inlet screen section may be inserted, and properly graded filter material may be placed in the annulus. Before placing the new screen, the old inlet should be heavily perforated to enlarge its open area. The old and new inlets should be fabricated from the same material to minimize corrosion caused by the introduction of the second inlet. Casing failures and faulty seals around telescoped screens can also admit sand into the well. The correction usually involves downhole repairs such as liners, connecting pipes of different diameters via swedging, and grouting.

Two commonly used methods of measuring sand content in well water are the Imhoff cone and the centrifugal sand sampler. The Imhoff cone holds 1,000 mL. One milliliter of sand is approximately equal to 1,000 mg/L (ppm) of sand by weight. A centrifugal sand sampler separates the sand from the water as the water flows through the device. This flow permits processing larger volumes of water more easily than can be done with an Imhoff cone to collect a significant volume of sand for better measurement accuracy. Other means of separating sand from pumped water, such as large centrifugal separators or a stilling well, may also provide means of measuring sand content. For recharge and ASR wells, suspended solids in the recharge water should be removed to the maximum extent possible. TSS can be measured efficiently with membrane or bypass (cartridge) filters, which are more accurate at low concentrations. As TSS frequently occurs as isolated events rather than continuous flows, bypass filters are preferred. Multiple membrane filters are also appropriate when use is for an extended period.

7.3.7 Site Security. Recharge facilities can be attractive to members of the public. O&M activities should include appropriate security procedures to protect the public from the potential hazards posed by recharge facilities. The security procedures should also protect the facilities from vandalism. Examples of security procedures include routinely inspecting facility premises; verifying that fences, buildings, and other structures that should be locked are so; and ensuring that surveillance equipment and security alarms are in proper working order. See also *Guidelines for the Physical Security of Water Utilities* (ASCE 2010).

7.3.8 Well Shutdown. Well shutdown, which is temporary, should not be confused with well closure (Chapter 8), which is permanent. Well shutdown is usually a relatively simple procedure. Activities should proceed in a general upstream to downstream direction to reduce the risk of encountering overpressurized components. All parameters should be carefully monitored, and any surges in flow should be noted. If surge protection was not included in the design of the initial recharge facility setup, it may be necessary to add this feature if significant surges are noted.

During well shutdown, operators should take advantage of the opportunity to check additional parts of the facility. Periodic shutdown of wells producing from sedimentary aquifers allows the backflow of water into the aquifer, where it may dislodge any fine sediments that had built up around the inlet section of the well casing and may carry them into the aquifer. Under this condition, it may be necessary to surge the well and waste the initial production when restarting the well.

Static water levels should be measured and recorded when the water level has recovered. An annual pump test at four different flow rates should be made to measure pump and well efficiency. Measurements should be made of power consumption, discharge pressure, and water level drawdown in the well for calculating input and output (water) power consumption. Specific capacity of the well (flow divided by drawdown) can also be calculated from these measurements and should be compared with previous values as an indicator of problems. Wire-to-water pump efficiency should be calculated and compared with the pump curve for indications of pump problems. Specific injectivity should also be periodically evaluated (recharge flow divided by increase in water level). Specific injectivity may change with time because of well clogging and also because of aquifer regional mounding.

7.4 WATER QUALITY MONITORING AND TREATMENT

Monitoring water quality and treating it before recharge and after recovery are essential activities for both surface and subsurface MAR operations to maintain efficient rates of recharge and to minimize adverse health and aesthetic impacts. Water quality monitoring includes both physical and chemical parameters. The parameters that affect MAR project goals should be treated to reduce those impacts, as discussed in Section 7.2.1 for surface facilities and Section 7.3.1 for subsurface facilities. Effective planning, feasibility studies, and pilot-scale studies should have identified potentially problematic water quality constituents, and the MAR facility should be designed to take those into account, as discussed in Chapters 5 and 9. However, variations in the quality of source water, in MAR operations, or in long-term effects may result in water quality impacts that could not be identified in predesign testing, so ongoing monitoring and treatment activities are necessary.

7.4.1 Monitoring. The objective of water quality monitoring is to understand how the quality of the water can affect both recharge facility performance and groundwater quality. For ASR wells, an objective is also to confirm that recovered water quality meets applicable standards. For recovery of stored water from ASR wells to meet drinking water standards, it is necessary to restore the chlorine or chloramine residual. The list of constituents and parameters monitored is site-specific and should be reevaluated periodically and updated if needed.

Water quality monitoring should be done to continually assess the impact of clogging on recharge rates. For physical clogging, monitoring of turbidity and TSS is important.

Monitoring for biological clogging mechanisms can include nutrients, temperature, and pH. Monitoring for chemical clogging mechanisms can be quite complex and should be defined after an evaluation by someone experienced in surface water, groundwater, and sediment chemistry. Particular attention should be paid to cases where the chemistry of the source water can vary. For example, if a recharge system has access to imported water, recycled water, or storm runoff, it is important that the potential of chemical clogging with all these sources and the sediments in the recharge facility be evaluated. The original feasibility study should address, among many other issues, any trends and variability in expected recharge water quality.

Temperature monitoring can also be important. Because of its effect on viscosity, warmer waters infiltrate at higher rates than colder waters. If there are large seasonal temperature swings, it may be necessary to adjust recharge facility performance to account for the effect of temperature. If the temperature of the source water exceeds that of the mean temperature of the upper 50 m (160 ft) of the receiving groundwater by more than 20 °C (68 °F), provision should be made to allow the surface water to cool before recharge operations proceed. This precaution is more important in well recharge but is also important in surface recharge because the impact on living organisms in the soil may diminish or destroy their beneficial cleansing action. For ASR wells storing drinking water at locations in northern latitudes, the seasonal variability in water temperature and viscosity, compared with typically much warmer temperatures in storage aquifers, can have a pronounced impact on the design, operation, and performance of pumps.

In addition to clogging, a water quality monitoring program should ensure that groundwater quality is protected. For example, in areas where the upstream watershed is being urbanized, stormwater runoff may contain heavy metals and other contaminants washed from paved or other improved surfaces. The concentration of these undesirable constituents may be greater in runoff caused by the first rains after a dry period. Such water could be rejected as being unacceptable for groundwater recharge. When recharge ponds are used for additional purposes, such as supporting a fishery, the effect of water quality and water treatment on the other purposes should be examined. An analogous situation for ASR wells is the runoff from initial snowmelt, which, in agricultural areas, can contain high concentrations of nitrate for several weeks in the late spring. At such times, recharge may need to be suspended. Recovery of stored water at such times can reduce or eliminate treatment costs for denitrification. This savings can be a secondary goal of ASR storage, following a primary goal such as meeting summer peak demands, drought, or emergency needs. A small increase in the target storage volume can provide for this additional ASR application, with no increase in capital cost for ASR facilities.

Groundwater quality in and around the recharge facility should be monitored. Even though filtration and soil aquifer treatment (SAT) are effective in improving the quality of the recharged water, groundwater quality monitoring should be conducted to ensure that the groundwater supplies are not being adversely affected by recharge activities. If the recharge facility is new, there may be a transitory change in groundwater quality caused by the flushing of salts and other constituents in the vadose zone. This phenomenon was observed in Arizona, where a new 95-hectare (236-acre) recharge facility was constructed in Avra Valley on farmland in 2008 by the Tucson Water Department. Within one month of recharge, a spike in nitrate concentrations was observed; however, within two years, nitrate concentrations dropped below background levels (see Case Study 5, Appendix D).

If there has been a history of recharge in the area, water quality constituents associated with previous recharge sources may also require monitoring. At the Montebello Forebay Spreading Grounds near Los Angeles, the historical use of wastewater for recharge has resulted in bound TOC in the vadose zone resulting from SAT. When advance-treated recycled water was recharged, the higher quality water had the potential to liberate the bound TOC into shallow groundwater (see Case Study 6, Appendix D). There is also the potential for leaching of naturally occurring constituents, such as arsenic, from the sediment below the recharge facility. Changes in local groundwater gradients caused by the recharge facility can also affect groundwater quality.

For ASR wells in deep, anoxic aquifers, recharge may mobilize arsenic and other metals because of the presence of oxygen in the recharge water. However, once the oxygen has been consumed because of geochemical reactions and microbial activity, the arsenic typically adsorbs and/or precipitates in the storage aquifer and is immobilized. This is often referred to as a roll-front deposit. For most ASR wells monitored to date, arsenic movement laterally within the storage aquifer is less than 60 m (200 ft). For ASR wells in shallower, unconfined aquifers for which the water contains oxygen, it is most likely that arsenic contained within arsenic-bearing minerals has long since been flushed from the aquifer.

7.4.2 Treatment. Three significant barriers to efficient recharge are bed load, suspended sediment (turbidity), and chemical constituents in the incoming water. The deposition of the coarser material classed as bed-load sediment decreases the storage volume of the ponds and the time between cleanings. Settling basins placed ahead of the recharge ponds capture the bed load and make removal and disposal easier.

The suspended sediment, primarily clays and silts that constitute turbidity, results in a rapid reduction in the recharge rate as the suspended material accumulates on the surface and near the subsurface of the recharge area. If excessive turbidity is present, a choice has to be made between frequent drying and cleaning to renew the basin recharge rates and reduction in turbidity before the water is introduced into the recharge area. The implementation of this latter choice results in operating criteria that set a limit on the turbidity of water to be recharged. The choice of criteria has to consider the availability of water, its turbidity, and the economics of renewing the recharge rate through cleaning and disposal of the accumulated silts and clays. Table 7-1 shows the suggested maximum particle size range in micrometers and the maximum particle concentration range in nephelometric turbidity units (NTU) and TSS for operational efficiency of recharge (ASCE 2001).

Turbidity is easily measured but is not as good an indicator as TSS. At best, correlation between the two is poor. TSS should be measured using membrane or bypass filters.

Treatment to remove the suspended solids, and thus reduce turbidity, does not, however, remove chemical compounds that may be in solution and should not be recharged. In areas where the upstream watershed is being urbanized, the runoff is likely to contain heavy metals and other contaminants washed from paved or other improved surfaces. The concentration of these undesirable constituents may be greater in runoff caused by the first rains after a dry period. Such water could be rejected as being unacceptable for groundwater recharge. When recharge ponds are used for additional purposes, such as supporting a fishery, the effect of water quality and water treatment on the other purposes should be examined. When the water to be recharged is imported and of different quality than local surface water, the possibility of

Table 7-1. Suggested Maximum Particle Size Ranges for Operational Efficiency.

Aquifer Type	Recharge Method	Particle Size (micrometers)	Maximum Particle Concentration	
			NTU	TSS (mg/L)
Alluvial	Basin or channel	100–500	5–10	
	Recharge well	<1–5		0–3
	ASR well	<1–5		0–5
Karst	Well recharge	100–500		0–5
	ASR	100–500		0–10
Fractured	Well recharge	100–300		0–5
Bedrock	ASR	100–300		0–5

adverse effects on the recharge rate because of chemical reaction between the imported water and the native soils should be evaluated. The outcome of such chemical reactions include sealing off the recharge area and loss of recharge potential.

Pretreatment of recharge water for wells may be required because of the possible presence of organic solids in the source water or recharge water piping. Such pretreatment may include bank filtration, sand filtration, settling basins, or other methods.

If the chemical quality of extracted or recovered water is deteriorating, some of the possible causes that should be investigated include the following:

- Leakage from nearby storage tanks;
- Leakage from landfills or other waste management facilities;
- Intrusion of saltwater or other poor-quality waters because of excessive drawdown;
- For ASR wells, recovery of water volumes substantially exceeding those for which the system is designed and operated (caused by recovering water from the buffer zone or the surrounding ambient groundwater, which is usually of poorer quality); and
- Recharge of inadequately treated waters.

As described in this section, understanding the physical and chemical composition of water that will be recharged and the treatment needed to make it compatible with native groundwater, are among the most important steps in determining the success of a MAR project.

CHAPTER 8
FACILITY CLOSURE

When permanently closing a MAR facility, care must be taken to close the facility in accordance with applicable local, state, and federal laws. If regulations require it, sediment and water samples should be collected in and around the recharge facilities and submitted for testing to determine if contaminants have accumulated in surface sediments, the vadose zone, and aquifer materials during the life of the recharge facility. If contaminants are found, a thorough analysis must be made to determine if they pose a threat to future land uses and groundwater quality.

8.1 SURFACE INFILTRATION FACILITIES

The amount and type of testing required before closing a surface recharge facility is, in part, dependent on the planned future use of the facility and associated property. For example, if the facility is going to be converted to a recreational lake, wildlife refuge, or park, then testing may be limited to measuring for regulated contaminants in the near-surface sediments to ensure that they are compatible with future uses. If the facility is going to be filled and used for residential or commercial purposes, then additional testing on the physical properties of the sediments in and around the facility may be needed. Particular attention should be paid to potential changes that may have occurred to the load-bearing capability of subsurface sediments, such as clays and silts, caused by interactions with the recharge water.

Any sediment in the bottom and sides of surface recharge facilities should be removed and disposed of in accordance with applicable regulations, if they constitute a threat to the underlying groundwater or to future uses of the site.

8.2 SUBSURFACE FACILITIES

Subsurface recharge facilities, such as wells or infiltration galleries, should be properly abandoned per regulatory requirements to limit the potential for leaving a vertical conduit that connects aquifer layers. Well abandonment can include ripping the casing and/or filling the well and space around the well with impermeable materials such as cement or clay, from the bottom of the hole up to the land surface (CDWR 1998).

Also, based on regulatory requirements, water quality testing may need to be conducted adjacent to subsurface recharge facilities to determine if residual contaminants have accumulated in the aquifer during the life of the recharge facility. If contaminants are found, a thorough analysis must be made to determine if they pose a threat to the water quality and what, if any, remedial actions may be necessary.

CHAPTER 9
DATA COLLECTION AND ANALYSIS

Many types of data are necessary to evaluate the planning and feasibility of a MAR project, design a system, track and evaluate its performance, and monitor its potential impacts. This chapter describes the parts of data collection (including organization and management), what data are needed (types and sources), how the data should be analyzed, the benefits of computer modeling, and its data requirements.

9.1 DATA COLLECTION

An experienced team of engineers, hydrogeologists, geochemists, and water quality specialists is needed to evaluate the technical feasibility and potential of a MAR project. MAR projects typically evolve through five principal phases—planning, design, construction, operation, and closure (Figures 1-1, 4-1, 5-1, 6-1, 7-1 and 9-1). Data collection for a MAR project evolves throughout the life of the project from high-volume data collection during planning and evaluation to less frequent and detailed data collection through construction, operation, and closure. The planning phase involves a prefeasibility phase that uses existing data and a feasibility phase that involves actual field data collection and may include small-scale pilot tests at one or multiple sites. The design and construction phases for a surface recharge facility focus on a single site, which should include a demonstration project that eventually evolves into a full-scale MAR facility. For a recharge well or ASR well program, demonstration testing at a single, full-scale ASR well is common. Results are scale dependent. Pilot testing with a small-diameter well is not recommended because results are not representative and may be disappointing, undermining support for continuation of the project. Demonstration testing at two or more wells is recommended because differences in well hydraulics and hydrogeology can reasonably be expected.

Figure 9-1 shows how the need for stakeholder confidence drives data collection in various phases and therefore time and cost. The challenge is to collect enough data at the right time to keep owners and stakeholders confident enough in project success to keep the project moving forward. Project owners assume more risk if they move forward without sufficient data. Project owners also may incur more cost if they collect data indiscriminately without identifying what data are most important for meeting project objectives (Chapter 4). With this goal in mind, sensitive parameters need to be identified and evaluated early to reduce the number of data collection iterations.

The process of data collection and analysis for a given project phase can follow a repeating circular path that builds the body of data and enhances stakeholder and owner certainty. A certain level of owner and stakeholder certainty is required to move a project to the next phase (Figure 4-1). Therefore, one or more data collection cycle iterations are required. Planners must identify data needs that should satisfy the necessary level of certainty that the objectives of a particular MAR project phase are being met. At that point, data needs are prioritized and then evaluated based on available time and budget. Data are then gathered (Sections 9.2 and 9.4), compiled (Section 9.3), and analyzed (Sections 9.5 and 9.6). During the analysis, data gaps are identified. After that round of data analysis, the owners and stakeholders reevaluate their level of confidence (Section 9.9) to proceed to the next project phase and may collect more data if significant data gaps and uncertainty exist. If time and budget are not available, the project may be delayed or determined infeasible. If time and budget are available, then additional data can be collected and analyzed. Figure 9-2 shows this iterative process.

9.2 DATA COLLECTION PHASES

The following subsections discuss the types of data that should be needed in each project phase. Depending on the particular project site, not every phase or data item discussed may be needed.

9.2.1 Planning Phase. Data collection forms the basis of the planning phase. The planning phase consists of prefeasibility (evaluate existing data) and feasibility (collect field data) data collection iterations (Section 9.4). During planning, several sites should be evaluated and compared. The data search should focus on landownership; regulatory and legal feasibility; and sensitive technical parameters such as geology, geochemistry, aquifer horizontal and vertical permeability, water levels, aquifer thickness, water quality and variability, storage properties, water supply availability and variability, water costs, and water demand trends and variability.

Additional fieldwork, detailed modeling, or small-scale pilot projects are sometimes necessary to reduce owner and stakeholder uncertainty about key project components before proceeding into the design and construction phases. If adequate and reliable input data are available, computer models (Section 9.6) should be used in the planning phase to identify and evaluate sensitive variables, prioritize data collection needs, evaluate pilot through full-scale designs, and estimate future operation and maintenance costs.

Prefeasibility Planning Phase. The prefeasibility or preliminary planning phase, as discussed in Section 4.3.1, typically does not require fieldwork. Existing geologic and hydrologic literature and readily available water resource data are sometimes enough to determine whether MAR is a viable means of increasing local water supplies. In many cases, the Internet can be a starting point to obtain valuable information on basic hydrogeology and existing projects within a region of interest. Often there are previously published reports prepared by governmental

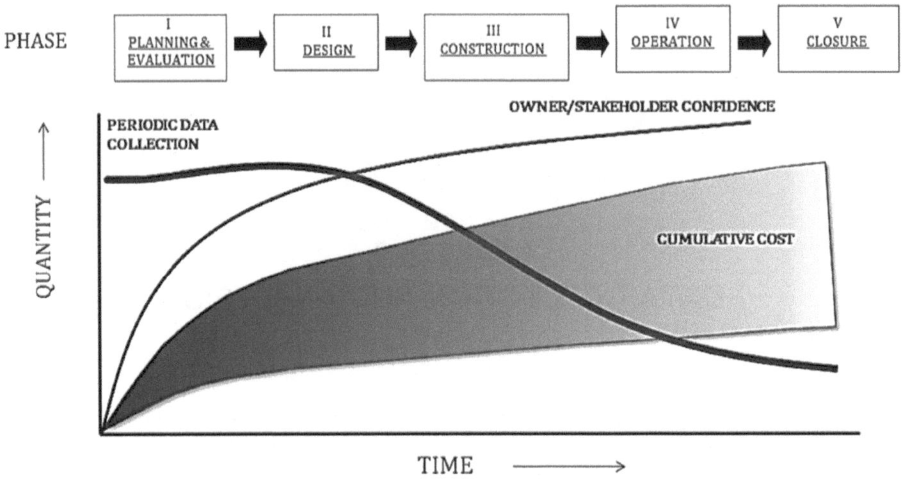

Figure 9-1. Data collection for MAR projects.

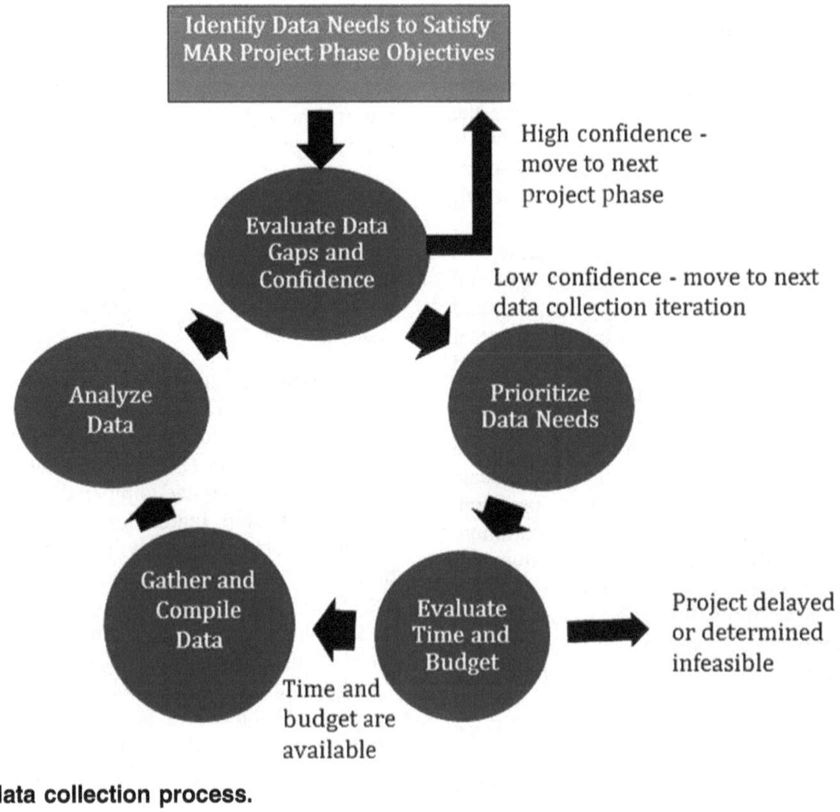

Figure 9-2. Iterative data collection process.

agencies like the US Geological Survey (USGS) or state geological surveys evaluating hydrogeological conditions pertinent to a project site.

In the prefeasibility planning phase, monitoring priorities are established that may include protecting existing land uses (wetlands, gravel pits), water uses and/or water rights, and environmentally sensitive areas (wetlands or existing contamination). Monitoring at existing wells and designated monitoring wells should be considered, and new monitoring well locations should be considered before implementing any fieldwork. Often, enough existing data are available for the MAR team to evaluate and rank technical alternatives (Section 4.3) and fieldwork options and to estimate preliminary exploration, infrastructure, and operational costs.

Feasibility Planning Phase. Once the decision is made to collect additional data, the extent of that data collection becomes a function of the owner's expectations for the MAR project. If the recharge needs are large, then the feasibility-level field exploration program may be extensive to evaluate several sites to levels that allow the owner to be relatively confident that the technology is feasible from a technical, regulatory, and legal perspective, enough to move forward with purchasing the site and committing to the design phase of the project. By the end of the feasibility planning phase, all but one site may be eliminated because of sparse existing data, existing contamination, proximity to sensitive areas, landownership, or an unfavorable geologic or regulatory climate. The culmination of the feasibility phase is usually a conceptual plan, which presents a comparative cost analysis of project options with recommendations for a single site moving forward to the design phase.

9.2.2 Design Phase. Chapter 5 discusses MAR project design and necessary data collection are discussed in detail. MAR project design is usually focused on one site based on planning-level data collection, analysis, and the results of pilot-level studies. Best management practices, including basin rehabilitation procedures and well backflushing practices (removal of particles that cause clogging), need to be defined and, if necessary, evaluated during limited-scale demonstration projects. If successful, a demonstration project can be expanded to the full-scale project in the operations phase.

9.2.3 Construction Phase. The primary objective of data collection in the construction phase (Chapter 6) is to create as-built drawings and surveys of measuring points and to establish baseline conditions, including depth to water. Some baseline performance-related data collection should occur, such as initial infiltration tests and pump testing. However, most data collection to evaluate performance occurs in the operations and maintenance phase (Chapter 7).

9.2.4 Operations and Maintenance Phase. During the operations and maintenance phase of a demonstration or full-scale project, system performance and monitoring data are collected and compared with the design plans to evaluate system performance and evaluate potential and predicted impacts. Changes in critical parameters need to be monitored, especially those relating to infiltration or well recharge rates, which can be easily affected by clogging. The degree of groundwater mounding needs to be monitored to assess the potential to influence local infrastructure, such as gravel pits, landfills, and underground structures, if present. Best management practices need to be reevaluated and refined on a constant basis because critical parameters, such as vertical permeability, saturated and unsaturated thickness, and well efficiencies, are dynamic throughout the life of a MAR project.

An example of monitoring in the demonstration phase of a project is the Bear Canyon Recharge Demonstration Project, in Albuquerque, New Mexico (Case Study 1 in Appendix D). The engineering team developed a monitoring program to evaluate the effectiveness of the project and to track the recharge water as it moved from land surface through the vadose zone to the underlying aquifer. A combination of temperature sensors, heat dissipation sensors, and neutron logging were used to monitor flux rates in the shallow [upper 15 m (50 ft)] vadose zone. Three monitoring wells were also used to regularly monitor water levels, collect water quality samples, and for neutron logging access through the deep vadose zone to the regional aquifer.

9.2.5 Closure Phase. During project closure, additional water level and water quality data are typically collected to validate model predictions and to continue to monitor sensitive areas. This MAR standard does not provide detailed guidance on data collection activities during closure, because post-closure monitoring requirements may be unique to the regulatory agencies that have jurisdiction for a particular project. Most likely, the same regional monitoring that occurred during operations continues during the closure phase of a project for some agreed-on period based on stakeholder needs to reduce uncertainty.

9.3 DATA ORGANIZATION AND MANAGEMENT

The collected data should be organized for easy retrieval and analysis. Where possible, automated data acquisition systems, such as supervisory control and data acquisition (SCADA) systems, should be implemented to compile time-series data into a database automatically. The database can be as simple as a spreadsheet. However, whenever possible, a spatial database, such as a geographic information system (GIS), is highly recommended for the storage and presentation of site-specific and regional data. The database can also be linked to a web-based server that allows owners and stakeholders access to data and customized analysis tools.

Careful thought should be applied to the database spatial coordinate system to take advantage of publicly available database systems and incur the least amount of error in database development.

9.4 DATA NEEDS FOR MAR FEASIBILITY STUDIES

Determining MAR feasibility involves collecting data on source water (Section 9.4.1); recharge method (Section 9.4.2); aquifer storage, transport, and recovery (Section 9.4.3); and potential impacts (Section 9.4.4). Table 9-1 shows the various data types needed to evaluate each MAR component.

Table 9-1. MAR Data Needs and Types for MAR Feasibility Studies.

MAR Component	Data Types
Source water	Water availability (quantity and timing)
	Trends and variability
Water demand	Trends and variability
Water quality	Trends and variability
Method (surface basin spreading or subsurface well injection)	Land availability and/or ownership
	Topographic information
	Groundwater levels (target aquifers)
	Surface infiltration rates and vadose zone properties
	Regulatory
Aquifer storage recovery	Geologic and geophysical data to define aquifer geometry and extents
	Aquifer properties
	Groundwater occurrence and flow data
	Groundwater quality data to determine compatibility with source water
Impacts	Regulatory
	Land use
	Vicinity well owners within the area of influence
	Regional water level data and monitoring
	Water rights
	Land subsidence

Several case studies are included in Appendix D. The Bear Canyon Recharge Demonstration Study (Case Study 1) provides detailed examples of data collection needs and analysis for site characterization, design, implementation, and permitting of an in-stream surface recharge facility in Albuquerque, New Mexico. The Rio Rancho Direct Injection Project (Case Study 10) includes similar information for the first direct recharge well in New Mexico.

9.4.1 Source Water

Water Availability. A MAR project is not feasible without a reliable water supply to recharge. Recharge water availability should be evaluated by a competent water engineer who should consider the following:

- Historical source water data (gauge data);
- Water costs (present and future);
- Physical water availability in the future;
- Timing of supply and recharge operations;
- Regulatory considerations (water rights and agreements);
- Existing storage, storage operations, and potential future needs (i.e., demands);
- Seasonal supply fluctuations and trends; and
- Climate change forecasts.

A source water delivery analysis is discussed in Section 9.5.1. If excessive uncertainty exists, additional data need to be collected or delivery modeling may need to be considered.

Water Rights Effects on Delivery. Freshwater is a scarce resource. A water right refers to the right of a user to use water from a water source such as a river, stream, pond, or source of groundwater. In the United States, each state has its own variations on these basic principles, as informed by custom, culture, geography, legislation, and case law. In areas with plentiful water and few users, such systems are generally not complicated or contentious. In arid areas, though, water rights are often the source of conflict, both legal and physical. Some systems treat surface water and groundwater in the same manner, whereas others use different principles for each.

Water rights play a vital role in the way deliveries may occur to a MAR facility. Some states have water rights models and decision support systems (DSSs) where the likelihood of water being available for a new facility can be evaluated. If water rights are an issue, an experienced water engineer should be a part of the MAR evaluation team to estimate water available (Section 9.5.1) in dry, wet, and average years.

9.4.2 Recharge Method. The recharge method can be through either surface or subsurface facilities. Determining which facilities to use involves collecting and analyzing the following data types:

- Landownership,
- Topographic data,
- Target aquifer characteristics,
- Water level data, and
- Vadose zone properties.

Landownership. Landownership is relevant for determining the recharge method because surface facilities have a significantly larger footprint than MAR well facilities. Facilities necessary for all MAR facilities include the recharge structures (basin and wells), conveyance facilities (pipelines and canals), and any surface storage (untreated settlement lagoons or treated water in storage tanks for subsurface recharge facilities). It is usually in the best interest of the MAR project owner to own the properties to have control over site access, security, and the development operation and maintenance of the facilities.

Site easements are frequently used for MAR well sites and pipelines. A MAR well typically requires less than an acre to accommodate the well, the well building, and electrical service, with adequate access for maintenance and repairs, plus regulatory setback distances from adjacent facilities such as sewer lines. Additional land may need to be owned, leased, or otherwise controlled to protect stored water from recovery by neighboring land owners. A wellhead protection area or well field protection area (WPA) may need to be negotiated with adjacent property owners surrounding recharge wells and ASR wells so that the water stored is available for recovery when needed.

Topographic Data. Topographic data often determine the recharge method because surface recharge facilities often require a large area with a sufficiently flat grade. Surface grading is a major cost consideration, and tilted areas require expensive containment berms.

Topographic data are also important for locating feeder canals, well infrastructure, pipelines, structures, and access. Topographic data come from existing topographic maps, land surveys, digital elevation maps (DEMs), or private specialty mapping, such as light detection and ranging (lidar) mapping.

A survey of MAR wellheads is also necessary to prepare water level elevation maps from measured water levels and to monitor future changes.

Target Aquifer Characteristics. Target aquifer characteristics are important for determining the recharge method. In general, all target aquifers should have the following characteristics:

- Be sufficiently large to provide adequate area and volume for MAR project storage;
- For surface recharge, have an adequate unsaturated thickness to allow the water table to rise without causing detrimental impacts;
- Have reasonably good aquifer properties (transmissivity, storativity, and for wells, leakance) so that water or pressure moves away from the point of recharge;
- Be able to contain recharge water in a way that can be recovered for later use; and
- Have water quality that will not hamper future uses of the recharged water.

Water Level Data. Water level data are necessary to determine the unsaturated thickness, determine the groundwater flow direction, and evaluate potential environmental impacts.

The depth to the water table or potentiometric surface is measured using a handheld device (electronic sounder) or by using an electronic data logger. The water table elevation or potentiometric surface is determined by subtracting the depth to water from the wellhead elevation. The wellhead elevation is determined using topographic maps or survey data.

Water levels are measured at existing wells or after drilling new wells. Calibrated instruments, such as well sounders and pressure transducers, are necessary to obtain accurate data. Measurement consistency is important to assess changes in water levels over time. Automated well measurement and data storage devices, such as transducers and data loggers, should be used whenever possible. See ASTM Standards D4750 and D4759 for procedures to measure water levels. Key considerations for using existing wells include the location of screened sections with respect to targeted aquifer zones, well construction materials, and the integrity of well completion so as to minimize the risk of cross-flow between formations and the ability of the well to be retrofitted for recharge.

Time series (multiple data measurements over time) measurements provide a useful indication of how the water table or potentiometric surface responds to temporal and seasonal variations in climate, aquifer recharge and discharge, and regional groundwater pumping. Time series data are sometimes available from federal and state agencies.

Vadose Zone Properties. The three-dimensional geometry and material properties of a target aquifer, including its vadose (unsaturated) zone, need to be evaluated to assess whether recharged water can be contained, controlled, and extracted as desired. The vadose zone should also be investigated to evaluate whether any significant restricting low-permeability layers exist and whether undesirable salts, chemicals, or other soluble constituents exist that can be leached out by infiltrating recharge water and carried to the underlying groundwater (see ASTM Standards D4696, D4700, D5126, and D5730; Bouwer and Rice 1984). Subsurface geologic features, such as fractures and faults, can control the flow of groundwater within adjacent strata. Potential connections with other formations that form other aquifers or confining barriers require careful delineation to make sure that the recharge water can be accounted for, controlled, and recovered. Subsurface characteristics can be evaluated through surface geophysics, test hole drilling, sampling and laboratory testing, and aquifer testing (Section 9.4.3).

In selecting sites for field tests and pilot projects, care should be exercised in the site selection so that results can be extrapolated with reasonable confidence to the entire area proposed for recharge. The area surrounding the test or pilot site should be well instrumented to record quantity and quality changes at frequent intervals using data loggers that transmit data automatically to a central processing facility or database, if possible. Ideally, field or pilot testing should be conducted for as long as possible and with the proposed water source that would be used for a full-scale project. Longer tests reduce the chance that results are skewed too high by the dry conditions and provide a sense of the magnitude of the clogging rate. In addition, using the proposed recharge source water, rather than a clean, potable source as is commonly used, provides additional data on the clogging rate.

For surface recharge facilities, the infiltration rate is one of the most critical parameters. Infiltration basin sites should have permeable soils (including sand–gravel mixes, sand, or loamy sand) but not too many cobbles because they can reduce infiltration rates and cause difficulties in basin rehabilitation. Data collected in evaluating the surface and near-surface sediments can include grain-size distribution analyses, acid residue analyses, X-ray diffraction analyses, porosity and grain-density measurements, specific gravity analyses, cation exchange capacity (CEC) with leachate analyses, and scanning electron microscopy (SEM). Grain-size distribution analyses are critically important in identifying zones that have high percentages of silt and clays (fines) that can restrict infiltration. The CEC analysis, in combination with knowledge of the recharge water chemistry, sheds light on whether there is the potential for exchange between the soils and recharge water. Cation exchange of clay minerals, such as calcium smectite (swelling clays), can cause the clays to disperse and migrate into the pore throats of the surrounding matrix, effectively reducing matrix permeability. In complex sites, it is important to bring in soil scientists and geochemists to ensure that there is a full understanding of potential interactions between the recharge water source and vadose zone soils.

Subsurface exploration should take into consideration geologic heterogeneity. It is usually necessary to drill test holes to determine if restricting layers exist and to obtain water quality and core samples of aquifers and interbedded layers at several depths. Permeability tests at various depths (ASTM D5084) are useful for identifying the effects of aquifer heterogeneity. Depth to groundwater should be assessed to make certain that storage capacity is available and evaluate the potential for perched groundwater, which is an indicator that a specific geologic layer is controlling the infiltration rate through the vadose zone. Therefore, aquifer heterogeneity can ultimately control the infiltration rate if water levels rise to the bottom of the infiltration basin.

Field tests should be conducted to evaluate and compare various hydrogeologic settings within a potential project area. For instance, a potential MAR site that has an ephemeral channel with a high percentage of cobbles may be less favorable for a recharge basin than an off-channel area situated within a dune sand deposit. These sites can be evaluated and compared by conducting standard infiltration tests (ASTM D3385). For example, trenches 1 m (3 ft) wide × 3 m (10 ft) long can be used to evaluate and compare different sites subjectively. A more formal pilot test can be done if there is confidence in the preceding work.

9.4.3 Aquifer Storage Recovery. To determine the amount of water that a MAR facility can recharge, store, and recover, the following data should be collected:

- Geologic and geophysical data to define aquifer geometry and extents,
- Aquifer properties,
- Water level data (Section 9.4.2), and
- Groundwater quality data to determine compatibility with source water.

Aquifer Geometry and Extent, Geologic Data. The aquifer geometry and extent should be delineated using existing geologic maps or through detailed surface mapping of bedrock types and physical properties. This information is critical to the understanding of natural groundwater flow conditions, specifically recharge and discharge, along with vulnerabilities to contamination from land use activities. Surface geologic exposures may also provide opportunity for detailed physical description and sample collection to define aquifer characteristics and evaluate geologic layering and heterogeneity.

The primary source of geologic data is from surface geologic mapping. Small-scale geologic maps prepared by the USGS may not be adequate to evaluate some MAR project sites, so larger scale geologic mapping may be required.

The extent of the target aquifer can be determined by identifying geologic formations that bound that aquifer. Boundary formations usually have a lower permeability (shale or clay formations), which frequently defines the bottom of the target aquifer. MAR facilities are typically located in the upgradient portion of the target aquifer to maximize capture of recharged water in downgradient extraction wells.

For ASR and recharge well projects in unconsolidated aquifers, and in some consolidated aquifers where there are geochemical concerns or uncertainty, continuous wireline cores are collected through the target aquifer storage interval(s), plus adjacent portions of the overlying and underlying confining layers. Cores are photographed and described on site by a geologist, after which selected core sections are sent to a core lab for detailed analysis. Core lab results are then evaluated relative to recharge water chemistry and native groundwater chemistry, using a geochemical simulation model to predict subsurface reactions and reaction rates. The results of the geochemical model, core lab results, and field geologist notes then provide a sound basis for development of pretreatment and/or well conditioning procedures. Use of drill cuttings in lieu of cores is often substituted because the cost is much less; however, the reliability of the resulting conclusions is also reduced, depending on the degree to which the drill cuttings are considered to be representative of the aquifer materials at any given well depth, plus the oxidation of the drill cuttings that occurs during collection and analysis.

Aquifer Geometry and Extent, Geophysical Data. Surface and downhole geophysical techniques have been used for decades in the mining and petroleum industries to evaluate subsurface geology. Hydrogeologists use many of these same techniques to investigate groundwater conditions. Early in a MAR project, surface geophysical techniques that are useful in determining depth to bedrock can be used to assist in siting good project locations and for identifying subsurface features that may warrant further investigation by test hole drilling, sampling, and logging. Geophysics conducted before and after a recharge cycle (see Section 9.4.3) can offer insights into how recharge water migrates through the vadose zone and/or the aquifer system. During the operation phase, geophysics can be used as a tool to assess the overall performance of the system compared with predictions. For example, subsurface downhole geophysical methods can add valuable detail that is not always discernible from lithologic logs of drill cuttings, such as subsurface layering; depth to bedrock; relative water quality; and locations of jointing, fractures, or faults.

The ultimate utility of individual geophysical techniques depends on the sensitivity of the technique relative to the scale of the investigation, whether the site is located in an urban area with structures or utilities that could interfere with the geophysical method, the amount of time and funding available, and the availability of other site data that can be used to calibrate and verify the geophysical results.

Surface Geophysical Methods. Numerous surface geophysical techniques can be useful in the evaluation of recharge sites (see for example, ASTM D5777). Some common techniques and their application to MAR project studies are described in Table 9-2 and in Chapter 7.

Subsurface Geophysical Methods. Subsurface geophysical methods are conducted in boreholes. They are used to identify features that may not be apparent from observation of drill cuttings collected during borehole drilling. Several different probes are usually lowered from the surface into an open boring; some can make continuous measurements throughout the borehole, and others rely on the presence of drilling fluids or groundwater (see ASTM D5753). Table 9-3 describes the most commonly used downhole geophysical methods and their application to MAR project studies.

Televiewer logs and radiometric logs are the only subsurface techniques that can be conducted in the unsaturated portions of boreholes without drilling fluid or groundwater. Natural gamma and temperature logs are the only subsurface techniques that can provide useful information in cased wells.

The quality of geophysical data is highly dependent on a number of factors, including the experience and skill of the field geophysicist; the condition and calibration of the geophysical equipment; natural site conditions such as geology, topography, and vegetation cover; access restrictions that limit full coverage of an area; the presence of buried or aboveground infrastructure, such as power lines, pipelines, and railroads; and, for some methods, even weather conditions such as passing electrical storms. As a result, geophysical data can be challenging to

Table 9-2. Surface Geophysical Methods.

Method	Properties Measured or Estimated	Utility of Method to MAR Project Studies
Frequency-domain and time-domain electromagnetic induction	Stratigraphic changes with depth	Identifies aquitard and aquifer boundaries both laterally and vertically over large areas
Gamma ray spectrometry (radiometrics)	Measures the presence of the radioactive elements potassium (K), uranium (U), and thorium (Th)	Delineates changes in depositional environments over large areas, helping to characterize regional stratigraphy
Magnetics	Measures the magnetic strength of rock, particularly rock types with iron-bearing minerals, such as certain volcanic rocks	Identifies depth to bedrock, basin-bounding shallow bedrock, or buried faults
Seismic refraction and reflection	Measures significant changes in seismic velocities of materials	Defines depth to bedrock; identifies presence of aquitards
Microgravity	Measures the change in gravitational acceleration caused by changes in the density of near-surface materials	Defines the geometry of the bedrock surface or the presence of significant buried faults
Controlled-source audio frequency magnetotellurics (CSAMT)	Identifies conductive and/or resistive zones in shallow alluvium or bedrock	Locates buried clay layers or maps the distribution of groundwater with a high specific conductance
Ground-penetrating radar	Identifies radar-reflective shallow subsurface features	Defines subsurface stratigraphy and identifies buried geologic structures

Table 9-3. Subsurface Geophysical Methods.

Logging Method	Properties Measured or Estimated	Utility of Method to MAR Project Studies
Spontaneous potential	Measures the electrical potential between depths in the borehole and a grounded voltage at the surface	Identifies coarse versus fine-grained layers, bed thickness, relative changes in groundwater quality
Normal and/or lateral resistivity	Measures the resistance of an electrical current passing through a formation	Delineates stratigraphic layers and determines bed thicknesses; identifies water-bearing zones
Natural gamma ray	Measures the natural radiation emitted by the minerals potassium-40, uranium-238, and thorium-232 within the formation	Identifies sandstone and clay beds; measures thickness of beds
Neutron	Measures the scattering of neutrons between a neutron source and a sensor	Determines the relative porosity of beds, moisture content of unsaturated formations
Gamma–gamma density	Measures the absorption of induced gamma radiation	Measures the density of formation materials
Elastic wave propagation (sonic logs)	Measures induced sonic wave velocities	Identifies formation consolidation, presence of clay beds, presence of fracture, and fracture density in consolidated rock
Acoustic and optical televiewer logs	Provides visual graphics of borehole wall using either acoustic reflections or optical instrumentation	Identifies formation contacts, fracture densities, and orientations
Temperature logs	Water temperature	Changes with depth can be used to determine vertical gradients, thermal gradients, fracture conduits, and lithologic changes
Flowmeter (spinner) logs	Measures the relative flow into or out of producing zones	Establishes the flow distribution downhole and how that changes with time because of well clogging, changes in head

interpret and, as such, other subsurface data should be collected to gain confidence in their interpretation.

Aquifer Properties. Several critical aquifer properties control the occurrence, movement, and potential recharge and withdrawal of groundwater. These critical hydraulic parameters for characterizing aquifers include the saturated and unsaturated thickness, aquifer hydraulic conductivity, transmissivity, storativity, and leakance through confining layers.

Often, existing data come from existing wells or studies (including federal or state mapping programs, contamination studies, geotechnical investigations, mineral exploration, and facility siting studies). It is important to assess how and why existing data were collected to evaluate their quality.

Of particular importance for ASR wells is the lateral rate of movement of the water in the storage aquifer, which depends upon the local and regional gradient and its variability, the hydraulic conductivity of the storage aquifer, and the bulk porosity. This issue should be addressed during the feasibility assessment stage of project implementation. For many ASR wells, the storage radius is quite small, such as a few hundred meters (feet), to achieve project goals. If the lateral movement of the stored water turns out to be several hundred meters (feet) per year, instead of a few meters (feet) per year, the stored water will move downgradient and not be recoverable from the ASR well when it is needed. This situation is partly why most ASR wells are in deep, confined or semiconfined aquifers for which the lateral rate of movement is typically quite slow. Such deeper aquifers often tend to have poorer water quality. Storage in deeper aquifers with slow water movement often requires use of thin, confined, and semiconfined storage intervals so that mixing with surrounding overlying and underlying poor-quality water can be effectively managed. The same principles apply to recharge wells; however, lateral movement away from recharge wells is accepted recovery is assumed to occur at a different well, potentially many months or years later, and blending with native water in the storage aquifer is also accepted. For many ASR wells, there is at least one water quality constituent in the surrounding storage aquifer that is not wanted in the recovered water. For marginal situations where lateral velocity is relatively high, ASR wells may be aligned in a row perpendicular to the average direction of groundwater flow. Recharge would be directed to upgradient wells with recovery from downgradient wells.

If one is conducting project-specific fieldwork and testing, ASTM standards are available to guide the data collection process and help to ensure data quality. ASTM Standards D5254, D5408, D5409, D5410, and D5474 describe the types of data that can be obtained from a groundwater site.

Collecting hydraulic conductivity, transmissivity, and storativity data should follow ASTM Standards D4043, D4044, D4050, D4104, D4105, D4106, D5084, D5269, D5408, D5472, and D5473. Soil properties (moisture content and porosity) can be determined using ASTM Standards D425, D653, and D5730.

Groundwater Quality and Compatibility. After evaluating the physical characteristics of a potential MAR project site to store recharged water, assessing the chemical compatibility is usually the next most important criterion when evaluating potential sites. Geochemical data and analyses are necessary to evaluate the potential for interactions between recharged water, soil, and groundwater. These reactions can lead to precipitation of minerals or development of biological activity that could clog

Table 9-4. Water Quality Parameters List for Geochemical Analysis.

Parameter	Parameter
Total alkalinity	Copper
Total dissolved solids	Manganese
Total suspended solids	Zinc
Turbidity	Cadmium
Color	Selenium
Specific conductance	Total hardness
pH (field)	Noncarbonate hardness
Temperature (field)	Calcium hardness
Dissolved oxygen	Nitrate
Eh (oxidation–reduction potential of the field)	Phosphate
Chloride	Ammonia
Fluoride	Hydrogen sulfide
Sulfate	Total organic carbon
Carbonate	Total halogenated hydrocarbons
Bicarbonate	Specific gravity or fluid density
Total silica	Total coliform
Calcium	Chloroform
Magnesium	Bromodichloromethane
Sodium	Dibromochloromethane
Potassium	Bromoform
Iron	Total trihalomethanes
Aluminum	

Source: Pyne (2005, Table 5.1).

the formation and decrease recharge rates. Mobilization of contaminants such as arsenic from soils containing pyrite is an important concern, as well as the chemical fouling of recharge and recovery wells by chemical precipitates. Laboratory analyses typically include a broad range of parameters (Pyne 2005), including those listed in Table 9-4. The expected value should be for each parameter. If a seasonal variation occurs then the annual range should be estimated. Doing so should help identify anomalous results and could indicate mobilization of dissolved ions.

Water quality data are typically tailored to the specific needs of each site. A few parameters that are monitored on a daily basis (Pyne 2005) may include the following:

- Electrical conductivity,
- pH,
- Temperature,
- Dissolved oxygen,
- Oxidation–reduction potential (Eh), and
- Turbidity.

Changes in these parameters over time, above natural or background variations, provide a general indication of mineralogical or biologic reactions that may be occurring as a result of MAR activities. Methods of evaluating and presenting water quality data can be found in ASTM Standards D5738, D5754, D5877, and D5903. The range and frequency of water quality sampling usually change during the test program. In addition, if treated water is recharged, concentrations of chloride and trihalomethanes in the recharged and recovered water should be monitored frequently.

Water Quality Databases. For many locations in the United States, regional water quality data can be obtained through the USEPA Storet database (https://www.epa.gov/waterdata/water-quality-data-wqx), the USGS National Water Information System database (http://qwwebservices.usgs.gov/), or through well sampling and laboratory analysis. Early in the feasibility phase, a complete suite of all USEPA priority pollutants should be collected from the target aquifer and evaluated to determine whether groundwater in the aquifer exceeds these criteria. In general, recharging a contaminated aquifer should be avoided owing to additional requirements that may be imposed on treating extracted water that may become contaminated.

Water Quality Concerns. Historical water quality data that are too old may not be representative of the current conditions in areas that have experienced significant growth or changes in basin activities, such as increased pumping stresses. Historic water quality analyses may have been taken during the investigation of a specific problem and therefore may be biased toward poor quality. In addition, historic data may be based on higher detection limits than current methods provide and so may bias potential contaminants as nondetect at the higher reporting limits. Registered monitoring wells may be indicators of areas of existing or past contamination. If a well owner that is a known industrial provider (such as a mining company or solvent manufacturer) or an oil and gas company, then additional information concerning the status of any cleanup activities may be available from the USEPA, the state oil and gas commission, the county health department, or the state department of environmental quality.

9.4.4 Impacts. MAR projects can potentially impact vicinity water users in the form of water levels, water rights, land subsidence and induced seismicity.

Regional Water Level Data and Monitoring. If a MAR project is going to be constructed, a groundwater level monitoring network should be established that is comprehensive and extensive enough to monitor environmental and stakeholder impacts. A monitoring network should extend beyond the area of expected changes in baseline water levels and water quality over the life of the project. The location of monitoring wells should consider the need for establishing baseline conditions so that a complete data set is available over the life of the project. Continuous data sets are useful for preventing litigation and if regulatory requirements change.

Baseline water levels, water quality, and unsaturated zone thickness need to be established in the feasibility phase across the entire MAR site and surrounding area of influence. Baseline water levels and quality may have changed significantly since predevelopment conditions, so care should be taken to interpret all available data to identify both existing water levels and current trends. If possible, baseline monitoring prior to construction should extend over several seasons, including agricultural growing and nongrowing seasons and urban peak and nonpeak pumping periods. Aquifer water levels are usually highest during the former, whereas water demands are highest during the latter. Baseline water level monitoring is done using existing wells or constructing dedicated monitoring wells to establish background conditions and monitor future changes. Careful consideration should be given when evaluating MAR sites where water levels are already increasing because of changes in water use or management, especially if the unsaturated zone is already thin. These conditions could reduce the overall volume of recharged water over the expected life of the project.

If wetlands may be created or adversely affected, baseline wetland surveys should be conducted. If land subsidence or uplift is a potential effect, then baseline elevation surveys should be implemented.

MAR Facility Effects on Other Water Rights. If water quantity, timing, or water quality impacts are significant, some states require augmentation or mitigation plans to be implemented before permitting a MAR facility. Recharge and withdrawals change groundwater hydraulic gradients in and out of streams that can significantly affect senior water rights. The timing and location of impacts to surface water right holders can be evaluated using analytical methods (Glover 1974, Jenkins 1970) and numerical groundwater methods (Section 9.6.2). Analytical distance–drawdown equations (Theis 1941) can be used to estimate water level changes and changes to flow gradients. Whenever hydraulic gradients change, it is possible that water quality impacts to water rights can occur. Numerical flow models (Section 9.6.2) and geochemical models (Section 9.6.6) can be used to predict water quality impacts caused by MAR activity. Evaluating MAR facility impacts on existing water rights should be evaluated in the prefeasibility phase because it can be a fatal flaw to a MAR project.

Land Subsidence and Heave. Changes in soil pore pressure can cause subsurface materials to heave or compact depending on the soil material properties, thickness, depth, saturation, degree of consolidation, and changes in pore pressure caused by MAR facility operation. Land subsidence occurs when fluid withdrawals reduce pore pressure or increase effective stresses, causing soils to compress and compact. Uplifting can occur when recharging groundwater increases the pore pressure or reduces effective stresses, resulting in elastic rebound of the formation (Miller and Shirzaei 2015, Pavelko et al. 2006, Zhang and Burbey 2016). Uplifting can be delayed months or years (Nikos et al. 2016). The mechanics of land subsidence and types of data collection and monitoring are described in ASTM D2435.

The Tonopah Desert Recharge Project (TDRP) is a surface recharge project located approximately 40 miles west of Phoenix, Arizona. The site began full-scale operations in 2006, and approximately 810,000 ac-ft of groundwater was recharged through 2014, when the project was discontinued because of lack of available recharge water. The TDRP is located adjacent to the Central Arizona Project (CAP) canal, which delivered up to 8.5 m^3/s (300 cfs) to the facility's 19 infiltration basins, totaling about 84 hectares (207 surface acres). Between 2006 and 2010, approximately 4 cm (1.6 in.) of uplift occurred, and maximum relative subsidence from November 17, 2007 to November 1, 2008, was about 1.4 cm (0.6 in.) (Budhu et al. 2014).

Satellite synthetic aperture radar (SAR) interferometry is ideally suited to measure the spatial extent and magnitude of surface deformation associated with aquifer-system compaction and uplifting. Techniques of interferometric synthetic aperture radar (INSAR)—principally coherent INSAR and to a lesser extent, persistent scatterer interferometry (PSI)—frequently are used to map, monitor, and analyze subsidence and uplifting associated with MAR. Figure 9-3 shows the location and extent of uplift resulting from MAR at the TDRP.

Induced Seismicity. Increases in pore pressure in seismically active areas can induce seismicity, depending on natural stress field orientation and the orientation of existing fractures. During the planning process, the status of a target area should be checked for seismic hazard using maps and data available through the USGS Earthquake Hazards Program (http://earthquake.usgs.gov/hazards/hazmaps/). Additional information can be obtained from detailed site geologic mapping, including borehole logging (Table 9-3) for the identification and characterization of fractures and faults in the target aquifer. It may be necessary to install local seismometers during the hydrogeologic characterization phase of planning to evaluate low-level seismic activity of an area. The USGS has instrumentation guidelines for seismometer operation (ANSS-WGD 2007).

9.5 DATA ANALYSIS

The purpose of collecting data is to evaluate critical processes for successful MAR activities, including water delivery, recharge, and impacts. The next sections describe numerous analysis methods. Evaluating data quality is critical during all phases of data collection and analysis.

9.5.1 Source Water Delivery Analysis. Source waters include storage reservoirs, rivers, canals, aquifers, and wastewater treatment plant effluent (Section 3.3). The delivery schedule is the rate or volume of water that is available for recharge and can vary widely. Deliveries from surface storage facilities are a function of the size of the outlet works or of the capacity of the pumping system. The availability of divertible river flows is highly variable depending on seasonality, controlled flow releases, daily rainfall events, and water rights considerations (Section 9.4.1). Canal deliveries are typically also seasonal and heavily influenced by upstream diversions. Effluent production at wastewater treatment plants can be steady or variable, depending on the treatment discharge stream. Source waters from aquifers may be from a different aquifer at the same location, or from the same aquifer at a different location. Both these MAR applications have been used for ASR well projects.

If deliveries are unpredictable or far in excess of recharge rates (such as during spring runoff), additional surface storage may be necessary. Surface storage, if economically feasible, could defeat the purpose of constructing a MAR facility. Therefore, a spreadsheet mass balance model (Section 9.6.1) can be used to evaluate sensitive variables that control source water delivery and how that can affect recharge deliveries and project viability.

Often the most cost-effective, viable project entails effective integration of surface storage and aquifer storage, taking advantage of the best attributes of each storage method. Surface storage reservoirs can capture runoff rapidly, but they are constrained by evaporation losses and adverse environmental impacts. They are also quite expensive and tie up large land areas. Aquifer storage can often store much larger volumes of water than surface reservoirs but can only recharge and recover at relatively slow rates. They are usually much less expensive than surface reservoirs and have essentially no adverse environmental effects. Combining the two water management approaches is often quite beneficial for achieving broad, regional water management goals. Surface water stored in reservoirs is transferred to aquifer storage, providing increased storage volume available aboveground to capture a greater portion of the subsequent runoff event.

Pretreatment Capacity. Suspended sediment needs to be removed from recharge influent to prevent excessive rehabilitation of recharge basins or to reduce frequency of backflushing of recharge wells. Natural variations in sediment load and treatment options need to be carefully evaluated to prevent surface recharge basin or recharge well clogging. Sedimentation basins provide time for suspended sediment to settle out before entering a recharge basin. Pretreatment by sedimentation is a function of flow rate, retention time, and therefore basin depth and area. Thus, seasonal source water sediment load and the dynamics and performance of sedimentation basin pretreatment should be carefully considered in system design. Again, because pretreatment can control actual recharge, it should be carefully evaluated in a spreadsheet model during the planning phase (Section 9.6.1) and more rigorously evaluated during the feasibility phase by conducting a pilot study.

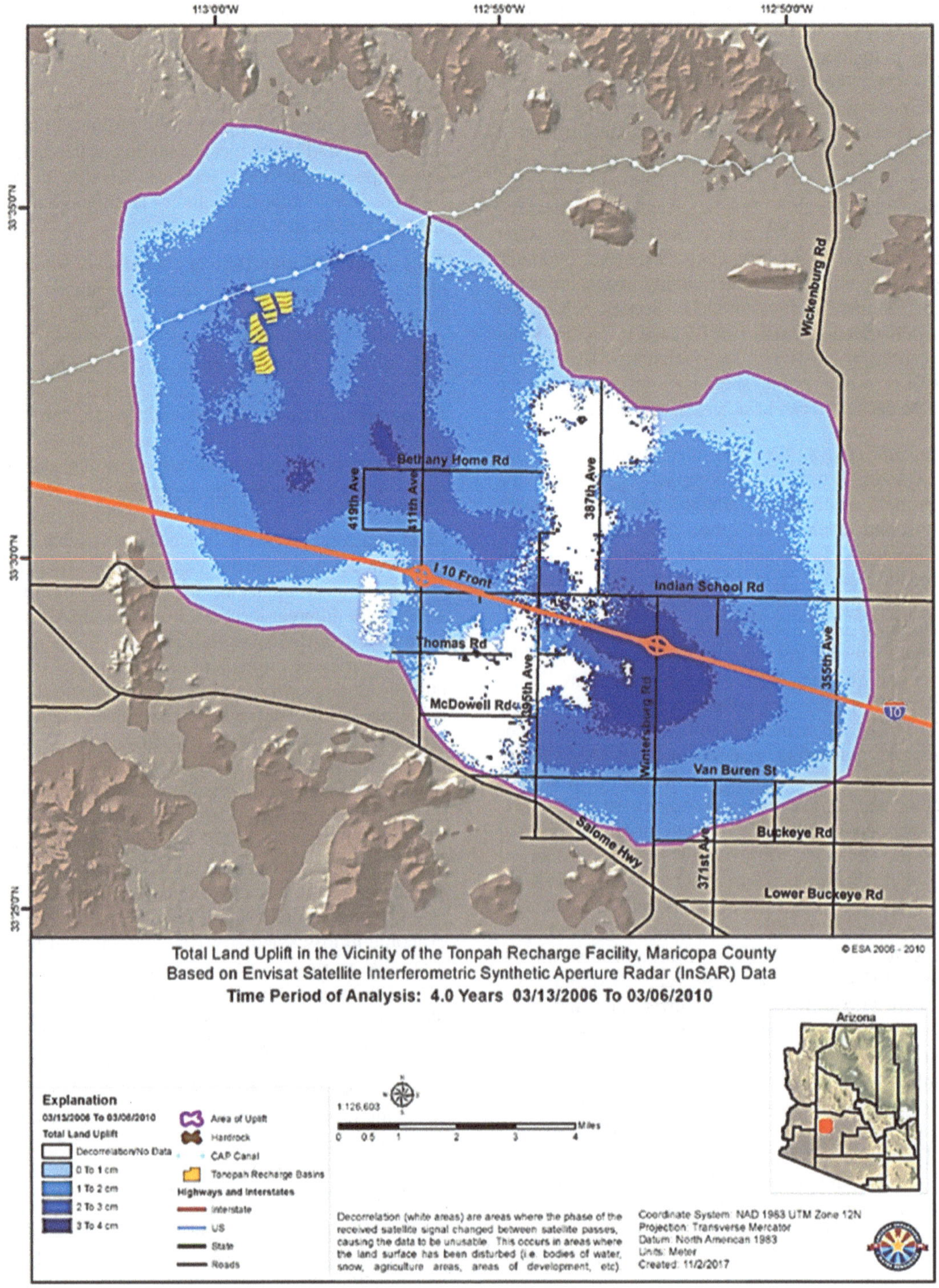

Figure 9-3. Uplift at Tonopah Desert Recharge Facility, near Phoenix.
Source: Arizona Department of Water Resources.

Infrastructure Rehabilitation. Spreading basins need to be routinely dried and treated to maintain infiltration rates and often to allow recharge mounds to recede below critical depths. Therefore, unless facility rehabilitation is planned carefully, recharge deliveries may be constrained by scheduled routine maintenance. MAR wells also need to be routinely backflushed to remove the buildup of suspended solids or geochemical precipitates, which reduce recharge inflow rates. These maintenance periods decrease overall recharge efficiency (Section 9.5.2). To compensate for infrastructure rehabilitation and other inefficiencies, additional spreading basins and MAR recharge wells may be necessary. Alternatively, higher quality water should be used for well recharge. Most ASR wells store treated drinking water.

9.5.2 Estimating Recharge and Recovery Volumes and Efficiency. During the planning phase, the amount of recharge and costs for various MAR design scenarios should be evaluated and compared on an equivalent cost basis. Costs include the sum of all infrastructure, engineering, and water costs over the project life determined on an average annual basis.

For ASR wells, costs should be prepared in terms of dollars per unit of volume, such as per million gallons or per acre-foot. Costs should also be prepared in terms of dollars per unit of production rate, such as cost per cubic meters (million gallons) per day (USD/MGD) of recovery capacity. Many ASR wells store relatively small volumes of water, sufficient to trim peak water demands for a few days or a few weeks per year. Annual volumes may be quite small, but the cost for meeting such peak demands with conventional treatment and conveyance facilities, instead of ASR wells, is typically substantially higher. Both economic measures are valid. To rely on one but not the other for decision making can lead to inappropriate conclusions.

Recharge volume considerations include the following:

- Delivery schedule;
- Basin dimensions, pond stage, and infiltration rate;
- Rates, volumes, diameter, and spacing of the recharge wells;
- Mound height and profile;
- Hydrochemical tracer methods (such as chloride balance);
- Wet/dry cycle times; and
- Recovery efficiency.

Once pilot testing and construction commence, these variables need to be compared with the actual design (Chapter 5) and future performance to establish operations and maintenance protocols (Chapter 7).

Surface Basin Dimensions, Infiltration Rates, and Volumes. The recharge volume (Q) in m^3 (af/day) for a MAR surface facility is proportional to the wetted area (A) in m^2 (ft^2) multiplied by the infiltration rate (I) in m/day (ft/day):

$$Q = A \times I$$

Theoretically, the larger the surface facility footprint, the more water can recharge. In practice, however, there is an optimal facility size that considers the delivery schedule and wet/dry cycle times.

The infiltration rate is an extremely sensitive variable that needs to be conservatively estimated during the planning phase (Chapter 4), evaluated during the design and construction phases (Chapters 5 and 6), and maintained during the operations phase of a project (Chapter 7). Once the vadose zone is fully wetted, the infiltration rate is a function of the hydraulic conductivity of the basin floor and the water depth (Bouwer and Rice 1989).

Actual basin performance should be evaluated during pilot testing and construction baseline tests (see Section 9.9) and compared with planning estimates. Figure 9-4 shows an example of infiltration rate versus basin depth data for a MAR surface recharge facility, which shows a strong correlation between infiltration rate and pond depth.

As suggested by Figure 9-4, recharge rates can vary considerably over a relatively small range of basin ponding depth. Therefore, this sensitivity should be thoroughly considered and evaluated during every phase of a MAR project.

Surface Recharge Mound Height. The height and rate that a recharge mound grows and moves away from a spreading basin is a function of the infiltration rate, unsaturated thickness, aquifer transmissivity, storage properties, and the physical boundaries of the aquifer.

Mound height calculations for canals (line sources) have been described by Glover (1960), and rectangular and round basin

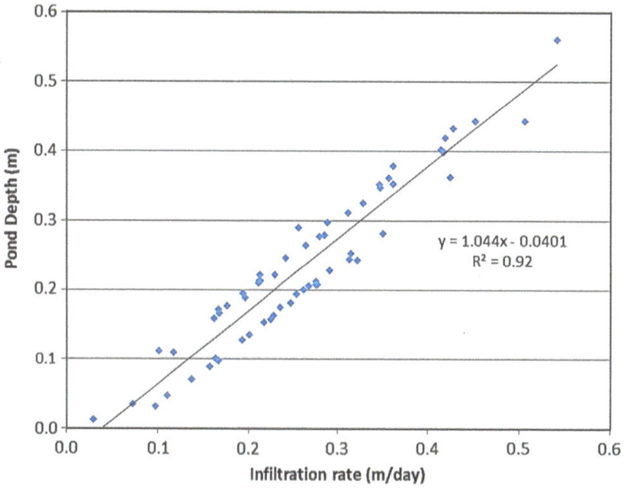

Figure 9-4. Infiltration rate versus pond depth.
Source: McGrane (2013).

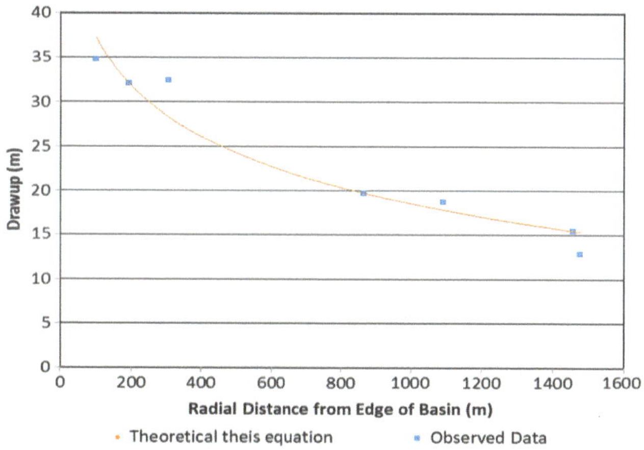

Figure 9-5. Mound height profile.
Source: McGrane (2013).

mound height calculations have been defined by Hantush (1967). Theoretical versus actual mound height profiling should be evaluated during pilot testing and construction baseline tests. Figure 9-5 shows theoretical versus actual mounding data obtained from monitoring wells at various distances from an infiltration gallery. The observations closely match theoretical predictions based on the Theis equation, using average aquifer parameters [transmissivity = 454 m^3/day (120,000 gal./day/ft) and specific yield of 0.05].

This back calculation of average aquifer parameters can then be used to refine future modeling predictions (Section 9.6.3) and monitor future project performance.

Well Recharge Rates, Volumes, and Spacing. For subsurface facilities, the mound height is described in terms of drawup in and around the recharge well. The recharge rate can be evaluated using variations of the Theis drawdown analysis (Theis 1952). A simple spreadsheet can be used to evaluate the sensitive variables, which include aquifer transmissivity, storativity, leakance, available drawup (distance from the potentiometric surface to the land surface), well efficiency, and time. Well diameter is a relatively insensitive parameter. Note that for most recharge wells and ASR wells, the potentiometric surface can rise to above the land surface during recharge, up to a pressure of about 200 kPa (30 psi).

Recharge well spacing is determined by using the Theis equation to maximize recharge rates and intermittent pumping for one or more recharge wells, while minimizing the well-to-well interference between them. In low-permeability aquifers, minimal well spacing can be hundreds of meters (yards). For storage aquifers that have poor water quality, well spacing is typically closer together, so that the recharge water can purge the aquifer of the native groundwater quality as the storage bubble expands.

Groundwater Flow, Hydrochemical Tracer Methods. Groundwater flow direction can be determined using water level and gradient data from monitoring wells. Actual migration paths and travel times can be more accurately determined by tracer tests. MAR tracer tests use water chemistry to track the source water from the recharge site to a downgradient monitoring point to evaluate groundwater migration paths and travel times. For ASR wells, the travel path and associated time of travel are particularly important, ensuring that the stored water can be recovered from the same well into which the water was recharged. Such a tracer can either be introduced separately into the groundwater system at the recharge site (added or extrinsic tracer) or occur naturally as a component of the source water being recharged (intrinsic tracer). Effective tracers are conservative (nonreactive), move in the same direction and at the same rate as groundwater (advective flow), and have concentrations or chemical signatures that are significantly different from ambient groundwater. Examples of added or intrinsic tracers include total dissolved solids (TDS), inorganic anions (including Br^-, Cl^-, F^-), trace gases (including SF_6 or He), dyes, or naturally occurring isotopes (Lee et al. 2009).

For an extrinsic tracer, dispersion and mixing in the aquifer can decrease tracer concentrations detected in downgradient monitoring points, requiring concentrations to be much higher than in ambient groundwater. For example, California regulations for recharge of recycled water currently require detection of an extrinsic tracer at a designated downgradient monitoring point of at least 10% of the original tracer concentration (SWRCB 2015). For intrinsic tracers, concentrations also need to be significantly different than in groundwater, but they could be higher or lower. For example, treated recharged water may contain much lower concentrations of certain constituents than occur in local groundwater, and the absence (or decrease) of such constituents would function as a tracer.

Overall water chemistry may also function as an intrinsic tracer when the hydrochemical signatures are sufficient to differentiate the source water from groundwater. Geochemical plotting techniques such as trilinear diagrams (Piper 1944) may assist in such an analysis. Naturally occurring isotopes can also be used as intrinsic tracers to evaluate migration and transport of source water in the groundwater system through groundwater age and estimated time of recharge. In the past decade, water isotope hydrology (particularly isotopes of ^{16}O, ^{17}O, ^{18}O, ^{1}H, ^{2}H, and ^{3}H) has become useful for determining groundwater sources, exchange of solutes (between groundwater and surface water), residence time, and flow paths (Kendall and Krabbenhoft 1995).

The travel time is typically evaluated from a breakthrough curve at a monitoring point some distance from the introduction of the tracer. The breakthrough curve plots tracer concentration with time and provides a means to identify the first arrival of the tracer, the peak concentration, the center of mass, and the trailing edge (Lee et al. 2009). Although there are a number of ways to interpret breakthrough curves for any particular site, the first arrival of the tracer is typically the best indicator of preferential flow and provides the most important timescale for evaluating retention times and transport of recharged water at a MAR facility (McDermott et al. 2008).

Simulating Wet/Dry Cycle Times. Spreading basins routinely clog because of sediment clogging the pore space or algal growth. Clogging rates can vary seasonally because of sediment load, temperature, and water quality. Dry cycles allow the basin to be rehabilitated by chemical treatment, or by physical scraping or tilling. The length of dry cycles varies seasonally depending on climate and soil conditions. It is important to determine reasonable wet/dry cycling periods during pilot or demonstration studies to determine recharge efficiencies so that project owners have realistic expectations concerning how much recharge will occur at their facilities.

Recovery Efficiency. The cumulative recovery efficiency (CRE) indicates the overall recovery efficiency of a MAR system for its operational or design life (Kimbler et al. 1975, Pavelic et al. 2002, Sheng et al. 2007, NRC 2008). The operational recovery efficiency (ORE) represents the efficiency for a single recharge–pumping cycle, as long as the recovered water meets target water quality criteria in terms of TDS or chloride concentration (Reese 2002, Pyne 2005).

Multiple hydrogeologic and operational factors control both CRE and ORE, including

- Losses to evapotranspiration (ET) by plants if a recharge mound height intersects the root zone,
- Discharge to surface water bodies,
- Losses to the target aquifer either through migration away from the recovery capture zone or as water retained in pore spaces (specific retention), and
- Losses caused by pumping by nearby nonproject wells.

Dry wells usually have low recovery efficiencies because they can be completed above the water table and typically do not have pumps for recovery. Recovery efficiencies increase if other dedicated recovery wells are used to recover recharged water.

Hydrogeologists, engineers, and soils scientists have developed both simplistic and sophisticated methods for estimating recharge and recovery efficiency components. Numerical groundwater models (Section 9.6.2) can more precisely estimate ET losses, discharges to surface water, and other factors listed earlier. However, accuracy is a function of the data and should not be confused with the sophistication of the analysis method.

Target Storage Volume. The target storage volume (TSV) is defined as "the sum of the stored water volume and the buffer zone volume in an ASR well" (Pyne 2005). If there is no significant difference in water quality between the recharge water and the native groundwater, buffer zone is not required and then the TSV is equal to the volume of water required for recovery to meet MAR program objectives. In most cases, this buffer zone is needed to ensure the quality of the recovered water. For most ASR wells, the storage aquifer contains at least one water quality constituent that is not wanted in the recovered water. Water needed to form the buffer zone is usually considered a "one-time" investment before normal MAR operations. During the MAR operation, additional water may be needed to augment the buffer zone periodically (Pyne 2005), particularly to make up for any lateral flow of the stored water. As a result, it should affect both operational and cumulative recovery efficiency. The volume of TSV depends partly on hydrological factors associated with the storage zone, such as thickness of aquifer, hydraulic characteristics, gradient, transmissivity, leakance, and water quality, and it also depends on MAR program objectives, such as recovered storage volume needs and duration of storage (Pyne 2005).

9.6 MODELING

A model is a more sophisticated method of data analysis that simulates the interactions of related variables. Models are developed with a purpose and are data driven. Models can be used to evaluate project planning and feasibility, design, operations, closure, and potential impacts. Planning models can be used to evaluate water availability and delivery schedules. Numerical groundwater models such as the analytical element models (Haitjema 1995) and the USGS MODFLOW program (Harbaugh et al. 2000) can be useful tools to evaluate MAR project feasibility and performance. Pre- and postprocessing programs enable complicated model inputs and outputs to be processed and visualized in a timely manner.

Modeling can reveal sensitive variables that influence project feasibility and drive data collection efforts. Models usually begin as spreadsheets and evolve to higher levels of complexity. Spreadsheet logic functions allow the interaction of several variables with linear relationships. Once too many variables are modeled or if their relationships become nonlinear, then a more sophisticated numerical groundwater model may be appropriate.

The most common form of a model is a spreadsheet mass balance model (Section 9.6.1), which allows the interaction of several variables normalized to a consistent time step. More sophisticated commercial probabilistic simulation software is also available for decision support and risk analysis. The International Groundwater Modeling Center (IGWMC), now called the Integrated Groundwater Modeling Center, was established in 1978 "to stimulate the correct and efficient use of simulation models and related computer-based support technology in the management and protection of groundwater resources." For more information on groundwater modeling, visit the IGWMC website at http://igwmc.mines.edu.

9.6.1 Mass Balance Models. Spreadsheet planning models usually are constructed using a mass balance structure where inflow, outflow, and storage terms are input and analyzed on a consistent time step. MAR project inflows include various source water delivery schedules and volumes, which are constrained by water rights, treatment, infiltration rate, mound height, and wet/dry cycling considerations. Outflow components include withdrawal schedules and losses caused by recharge and recovery inefficiencies (Section 9.5.2). The calculated change in storage reflects predicted water level changes over time.

Once a spreadsheet hydrogeologic model is created, it can be used to design and simulate MAR well field operations. For example, 114 m^3/h (500 gpm) may be available four days each month for injection. The modeler can calculate the drawup and well-to-well interference of a simplified square well field arrangement using four equally spaced wells injecting at 28.4 m^3/h (125 gpm). If the unsaturated thickness at the well field is 70 m (200 ft) and the well efficiency is high, the modeler may determine that it only takes two days to simulate the head buildup of 70 m (200 ft). If so, then the number of wells may be decreased because more drawup can be expected if pumping twice as long. As an alternative scenario, the planner may reduce the well spacing (which increases well-to-well interference and drawup). Under the alternative scenario, the decreased pipeline cost between MAR wells may be more cost effective than installing fewer wells. This example demonstrates how spreadsheet models can be used to evaluate well field design, operations, and cost.

9.6.2 Numerical Models. Numerical groundwater models, such as the USGS MODFLOW program, allow the simulation of more complex hydrogeologic settings and more sophisticated sensitivity and uncertainty analyses and probabilistic determinations. MODFLOW is a modular program that can incorporate a wide variety of specialized packages. It allows increased sophistication depending on data availability and purpose and is widely used by both practitioners and regulatory agencies to predict groundwater changes induced by either pumping or recharge. MODFLOW is incorporated in various graphical user interface (GUI) programs that simplify model development, modifications to the model, and the processing of model output.

Numerical models are commonly used to evaluate the following:

- Data adequacy and needs,
- Parameter sensitivity and uncertainty,
- System performance,
- Operational alternatives,
- Future monitoring and verification,
- Water quality changes, and
- Fate and transport of mobilized contamination.

Simple numerical groundwater models consider important aquifer boundary conditions and variable delivery schedules. Sophisticated numerical models can include multiple aquifer layers, aquifer heterogeneity, complex boundary conditions, anisotropy (variations in hydrogeologic parameters), the influence of surface water features such as lakes or streams, and alternative delivery options (including average year, dry year, or wet year). Coupled surface water and groundwater numerical models have the highest complexity. Solute transport models such as MT3DMS (Zheng 1990) can be developed to assess the fate and transport of potential contaminants in the vicinity of a MAR site. Many guidance documents have been developed for various aspects of developing, evaluating, and documenting a numerical groundwater model, including ASTM Standards D5447, D5490, D5609, D5610, D5611, and D5718; Anderson et al. (2015); and Reilly and Harbaugh (2004).

MODFLOW modeling was used to evaluate the timing of well pumping depletions and surface recharge pond accretions for the Tamarack Ranch Recharge Project in Colorado (see Case Study 4 in Appendix D). The Tamarack Ranch multipurpose facility is managed by three states (Colorado, Wyoming, and Nebraska) and the US Department of the Interior. The program involves pumping 17 wells during the winter low-demand period, surface recharge in 11 ponds totaling 32 hectares (80 ac), and timing the accretions back to the stream in the summer to augment flow to benefit multiple endangered species. Colorado water users can also lease excess accretion credits for private augmentation plans. A MODFLOW model developed by the state of Colorado was used to locate extraction wells and recharge ponds and to support numerous water rights cases.

9.6.3 Data Collection for Modeling. The quality of model output is directly proportional to the quality of the model input data. Calibration is the process where model parameters are adjusted so that model output best matches real-world observations (ASTM D5981). Hydrogeologic data parameters, such as infiltration rates, horizontal and vertical hydraulic conductivities, and storativity, are modified during model calibration to match observations. The better a model compares to observed data, the more confidence there is in its model predictions.

Spreadsheet and numerical models are calibrated to the widely varying data, including infiltration rates and volumes, aquifer test parameters, water level time series, and water quality data. This information needs to be meticulously measured and recorded. Data recording is often done using continuous data loggers or SCADA systems.

Hydrogeologic data are usually measured at discrete points, such as within boreholes or wells, or average properties measured between observation points. Measured data should be gathered beyond the MAR project's area of influence or extrapolated with reasonable confidence. The MAR design team needs to prepare maps showing the spatial extent of critical hydraulic parameters, including aquifer hydraulic conductivity, saturated and unsaturated aquifer thickness, and transmissivity. Spatial data can be extrapolated manually using professional judgment, using gridding and contouring programs, or using inverse modeling techniques such as PEST (Doherty 2005) or UCODE (Poeter et al. 2005). These programs automatically evaluate parameter sensitivity and adjust spatial parameter values so that model predictions best match historical observations.

Temporal data are also important to model calibration. The best calibrated models can be shown to simulate changes in water levels over time by comparing changes in water levels measured at individual points with model predictions at those same locations. Therefore, water level data over a long period of time is important to acquire both at a potential recharge site and in the area around the recharge site. Figure 9-6 shows observed versus modeled water level elevations on a daily time step. The data were used to evaluate aquifer parameters and to determine how much more recharge may be possible to maintain water levels below the point of compliance of 312 m (1,024 ft).

If water level data show significant changes over time, it is important to understand the underlying cause of the rise or fall of water levels so that the model can incorporate the process that is causing the water level change. For instance, if wells within a model domain exhibit rises over a period of time because of wet climatic conditions, recharge to the model should be increased sufficiently to cause the model-predicted water level to rise. The resulting model is much more defensible if the cause and effect can be demonstrated and simulated in the model.

9.6.4 Determining Sensitive Variables. Quantifying parameter sensitivity is important for evaluating project technical feasibility, estimating cost, and prioritizing future field investigation and testing programs. Data collection should focus on variables and parameters to which model predictions are sensitive.

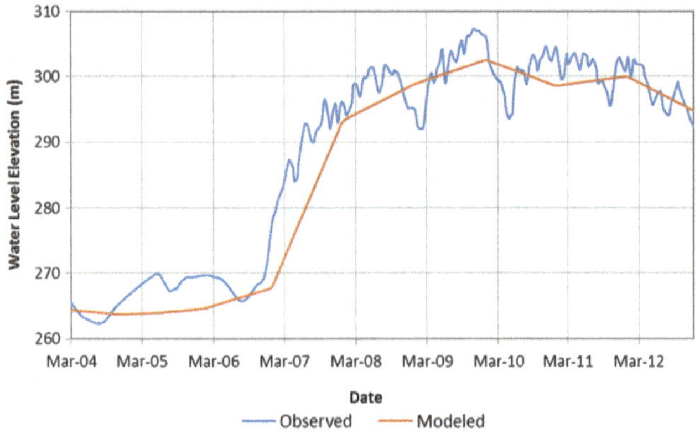

Figure 9-6. Modeled versus observed water levels at point of compliance.
Source: McGrane (2013).

Sensitivity analyses can be done using simple analytical equations described herein over a range of parameter inputs and comparing the results. Simple spreadsheet calculations can be used to evaluate the sensitivity of critical variables (such as rectangular area, lengths and widths, and depth) or to calculate mound height based on a range of delivery schedules, infiltration rates, and wet/dry schedules (Section 9.6.2). For example, the success of an infiltration basin generally depends on the vertical permeability of the unsaturated zone and the transmissivity of the aquifer. Therefore, the success of the project depends on whether there are any low-permeability clays interbedded below basin infiltration test locations. As a result, test hole drilling, sampling, geophysical logging, and infiltration testing are likely to be more important for evaluating project success than measuring water levels or sampling groundwater quality. This type of sensitivity analysis can be automated using inverse modeling programs, such as PEST (Doherty 2005) and UCODE (Poeter et al. 2005). Sensitivity analysis results should be used to evaluate data gaps and to prioritize future data collection efforts.

9.6.5 Data Collection for Uncertainty Analysis. Groundwater modeling is an uncertain science. Errors can exist as geologic, parametric, and predictive uncertainty, and models necessarily simplify complex natural systems to allow computer analysis. Each phase of the project has an increasing level of model detail, model quality, and uncertainty analysis that is appropriate for the amount of data available and the current goals of the project. The variability of aquifer properties combined with the difficulties and expense of data collection lend cause to extrapolating data and assumptions based on scientific expertise. These assumptions can be tested and supported by advanced computer techniques PEST (Doherty 2005) and UCODE (Poeter et al. 2005) that relate real-world observations to computer model calculations and can provide a measure of model prediction uncertainty. This uncertainty can help prioritize future data collection activities (see ASTM Standards D5447, D5490, D5611, and D5981).

Initial modeling of ASR well field predicted performance is often constrained by lack of data because aquifers selected for storage are often deeper than those for which local data are readily available. Leakance of confining layers is important for ASR operations, and such data are usually not available at the start of a project. Coring and construction of deep monitor wells early during project development help to gather the missing data, particularly including the confining layer underlying the storage aquifer. Construction and testing of demonstration ASR wells then achieve further data refinement, providing a firmer basis for conclusions regarding anticipated ASR performance, calibrated against actual performance of the demonstration ASR well during the first year of operations. Design of the ASR well field expansion can then be implemented with greater confidence, based on the performance of the demonstration ASR wells and associated simulation modeling.

9.6.6 Geochemical Modeling. Adverse geochemical interactions can be a fatal flaw to a MAR program because of clogging of aquifer or vadose zone soils. Several computer programs are available to evaluate different proportions of recharge and native water and to predict whether precipitation or dissolution reactions should occur. Chemical precipitation can cause aquifer clogging, and chemical dissolution can mobilize toxic ions that can be a health hazard. In some cases, MAR facilities mobilize contaminants such as nitrate or arsenic that were either introduced by prior land use (such as agricultural practices) or naturally occur in the sediments or rock comprising the aquifer. Therefore, the water interaction with native aquifer water and formation materials needs to be assessed.

Geochemical models that should be considered include the USGS's water equilibrium (WATEQ) model and pH redox equilibrium equations (PHREEQE) models; the USEPA's MINTEQ model; Lawrence Livermore National Laboratory's EQ3NR/EQ6 model; and The Geochemist's Workbench software (Pyne 2005).

9.7 MONITORING AND COMPLIANCE DATA

Regulators and other stakeholders heavily influence monitoring parameters, locations, and frequency (see ASTM D4050) during each MAR project phase. To reduce data collection iterations, data collection and monitoring requirements should be considered simultaneously. Meeting with regulators and stakeholders early and frequently is likely to reduce project development and permitting time and overall project costs. Monitoring requirements are typically substantial during early operational testing, declining significantly after satisfactory performance has been confirmed and operating permits have been issued.

9.8 OPERATIONAL DATA COLLECTION

Startup and operational data include the following:

- Flow rate, duration, cumulative volume, and quality of source water;
- Water balance accounting, including inflow rates, duration, and quality of the recharge to various facilities (wells and spreading basins);
- Flow rates, duration, and quality of water recovered from a MAR facility;
- Changes in wellhead pressure;
- Depth to and quality of groundwater under the recharge and adjacent areas;
- Power use by units and cumulative total;
- If nonpressurized flow in recharge wells, depth to water in the well;
- If pressurized flow in recharge or ASR wells, wellhead pressure;
- Precipitation on and evaporation from surface ponds;
- Temperature of water at inflow and outflow locations;
- Time, rate, and volume of pumping for units and total system;
- Water level response in observation wells;
- Wetland extents;
- Monitored parameters in environmentally sensitive areas;
- Equipment condition; and
- Any other data required by regulators.

Sections 7.2.1 and 7.3.1 discuss how record keeping is key to determining how well MAR facilities are being operated. Operational forms are typically designed by the owner, operator, or regulator. Time series trends of the aforementioned data sets should be analyzed in a timely manner to make operational changes that optimize system performance. The frequency of data collection may vary depending on the parameter and time since start-up. For instance, recharge rates decrease with time, so more frequent basin stage measurements or recharge well water levels are required immediately after start-up compared with later in an operations cycle.

9.9 DATA COLLECTION TO EVALUATE PERFORMANCE

MAR facility performance data should be collected during pilot studies, after construction (Chapter 6), and during the operations phases to be compared with the design (Chapter 5) and then used to evaluate future performance and maintenance procedures (Chapter 7).

Performance parameters and relationships for surface MAR facilities include

- Recharge rate and volume versus basin stage and/or depth (Figure 9-3),
- Recharge rate and volume versus wetted acres,
- Quality of recharge water (temperature, total suspended solids, turbidity),
- Wet/dry ratios [percentage of time a basin is being used (wet) versus time being dried for treatment], and
- Mound height versus time and distance relationships (Figure 9-4).

Performance parameters for subsurface MAR facilities include

- Recharge or ASR well drawup at similar recharge flow rates versus backflush frequency or volume;
- Well backflush procedure, frequency, and duration versus time;
- Recharge volumes versus backflush frequency;
- Recharge cycle efficiency [(total volume recharged – backflush volume)/total volume] versus time;
- Cumulative recharge efficiency;
- Total monthly and cumulative recharge volume versus time; and
- Mound height profile (various distances) versus time (Figure 9-5).

Any deviations from anticipated performance should be evaluated, diagnosed, and resolved before initiating full-scale operations. Once baseline performance relationships are established, they should be compared with operations performance to evaluate the need and effectiveness of maintenance procedures (Chapter 7).

9.10 REPORTING

Technical reports are usually necessary throughout a MAR project. Reports typically communicate data, analyses, conclusions, risk, and recommendations to the project owner and stakeholders during each MAR project phase. Data reports may form stand-alone documents specific to the type of data collected, analysis (such as basin or recharge well performance), or modeling effort, or they can be parts of or appendixes to broader planning documents such as conceptual plan reports. They should be complete and detailed enough so that data can be validated and that efforts can be recreated with data presented.

APPENDIX A
GLOSSARY OF GROUNDWATER TERMS

Alluvium: A geological term describing beds of sand, gravel, silt, and clay deposited by flowing water.

Appropriative Right: A concept in water law under which users who demonstrate earlier use of water from a particular source are said to have rights over all later users of water from the same source.

Aquiclude: A formation that, although porous and capable of absorbing water, does not transmit it at rates sufficient to furnish an appreciable supply for a well or spring.

Aquifer: A water-bearing soil or rock that contains sufficient permeable material capable of yielding significant quantities of water to wells and springs.

Aquifer Storage and Recovery (ASR): Storage of water in a suitable aquifer in a surface spreading basin, vadose well, or saturated zone well (ASR well) when the water is available, and recovery of the water during times when it is needed from a well or other means. This term is similar in meaning to managed aquifer recharge (MAR) as used in this standard.

Aquifer Storage and Recovery (ASR) Well: This is a dual-purpose (dual-function) well that is used for both recharging water into an aquifer and pumping (recovering) the stored water from the aquifer and is commonly designed for seasonal or long-term storage of water in an aquifer system. The well is equipped with a pump to permit recovery of stored water and periodic backflushing to remove accumulated solids and control clogging.

Aquitard: A soil or rock that is lower in permeability than adjacent deposits that retards but does not prevent the flow of water to or from an adjacent aquifer; a leaky confining bed. It does not readily yield water to wells or springs, but it may serve as a storage unit for groundwater.

Area of Influence of a Well: The area surrounding a well within which the potentiometric surface has been lowered; may be a transient or steady-state condition, depending on volume and duration of pumping.

Arroyo: An arroyo is an almost vertically walled, flat-floored stream channel that forms in fine, cohesive, easily eroded material. Arroyos exist throughout the western United States but are most common in arid and semiarid climates in the Southwest.

Artesian Well: A well deriving its water from an artesian or confined water body. The water level in an artesian well stands above the top of the artesian water body it taps. If the water level in an artesian well stands above the land surface, the well is a flowing artesian well.

Artificial Recharge: The deliberate replenishment of the groundwater supply by means of spreading basins, recharge wells, irrigation, or other means to induce infiltration of surface water. May include the substitution of surface water deliveries to water users who would otherwise extract groundwater, thereby leaving water underground (this is the "in lieu" method). This term has been replaced with managed aquifer recharge; *see* Recharge.

Available Storage Capacity: The volume of the total storage capacity that does not presently contain groundwater and is therefore available to store recharged water.

Bank Storage: The storage of water in an aquifer adjacent to and interconnecting with a surface water body so that a change in a stage of the adjacent surface water body causes a change in storage of water in the aquifer.

Buffer Zone (BZ): The recharge water volume that is initially formed and then maintained around the target storage volume.

Capillarity: The rise or movement of water in the interstices of a soil or rock caused by capillary forces associated with the fluid property of surface tension.

Capillary Fringe: The lower subdivision of the unsaturated zone immediately above the water table in which the interstices are filled with water under pressure less than that of the atmosphere. Being continuous with the water below the water table but held above it by surface tension, its upper boundary with the intermediate belt of the unsaturated zone is indistinct.

Capture Zone: The specific flowpaths that define the three-dimensional, volumetric portion of a groundwater flow field that provides groundwater to a pumped well.

Clay: (1) A fine-grained geologic material [grain size less than 0.004 mm (0.0002 in.) in diameter], which has low permeability, or (2) sedimentary deposits composed primarily of these particles.

Coefficient of Permeability: *See* Hydraulic Conductivity.

Cone of Depression: A depression in the water table or potentiometric surface of a groundwater body that is in the shape of an inverted cone and develops around a well that is being pumped. It defines the area of influence of the pumping well.

Confined Aquifer: An aquifer bounded above and below by beds of distinctly lower permeability than that of the aquifer itself, and whose water levels rise above the top of the aquifer layer; an aquifer containing confined groundwater.

Confining Bed: A body of impermeable or distinctly less permeable material stratigraphically adjacent to one or more aquifers.

Conjunctive Use: A term used to describe operation of a groundwater basin in coordination with a surface water reservoir system.

Consumptive Use: The total water loss from a water supply or system by evaporation or transpiration from a vegetated or non-vegetated surface, commercial or industrial process, and all domestic and municipal uses. The difference between the quantity of water withdrawn from a source and the quantity returned to the source or another source of usable water.

Cumulative Recovery Efficiency (CRE): The efficiency of a single recharge–pumping cycle, defined as the volume of recovered water divided by the recharge volume, expressed as a percentage.

Darcy's Law: An empirical law based on experimental evidence for the flow of fluids with the assumption that the flow is laminar and that inertia can be neglected. It states that the velocity of the flow through a formation is directly proportional to the hydraulic gradient.

Deep Percolation: The drainage of soil water downward by gravity below the maximum effective depth of the root zone toward storage in subsurface strata.

Drawdown: The lowering of the water table of an unconfined aquifer or of the potentiometric surface of a confined aquifer, usually caused by the pumping of groundwater from wells.

Dry Well: *See* Vadose Zone (Dry) Well.

Eductor Lines: Small-diameter pipes consisting of solid casing placed in a well below the saturated level. Are used to add recharge water to the well to reduce air entrainment compared with adding recharge water at the top of the well.

Effective Hydraulic Conductivity: The rate of flow of water through a porous medium that contains more than one fluid, such as water and air in the unsaturated zone. It should be specified in terms of both the fluid type and content and the existing pressure.

Effective Porosity: The amount of interconnected pore space available for fluid transmission. It is expressed as a percentage of the total volume occupied by the interconnecting interstices.

Elevation Head: The head that corresponds to the potential energy of a liquid, measured relative to a datum such as mean sea level.

Equipotential Line: The trace of a line connecting points of equal potential on a potentiometric surface or a two-dimensional cross section.

Evaporation: The physical process by which a liquid, such as water, in a stream, lake, or moist soil is transformed into a gaseous state. It may be expressed as the total or the mean rate in units of mass or volume per unit area, or as an equivalent depth of water for the period concerned.

Evapotranspiration: The total water vapor loss from an area by evaporation and transpiration from plants over a given time period. It includes the evaporation of water from soil, from dew, and from intercepted precipitation, and transpiration from plants.

Fault: A fracture in the Earth's crust, with displacement of material on one side of the fracture with respect to the other. Frequently acts as a barrier to movement of groundwater, but in some situations may act as a conduit.

Fine-Grained: (1) Said of a crystalline rock in which the individual minerals have an average diameter of less than 1 mm (0.04 in.), or (2) said of a soil in which silt or clay predominates.

Formation: A geologic term that designates a specific group of beds of strata that have been deposited in sequence one above the other and during a specific period of geologic time.

Free Water Elevation: The elevation at which pressure in the water is zero with respect to atmospheric pressure. Also termed groundwater surface, groundwater elevation, free water surface, phreatic surface, and water table.

Free Water Surface: The elevation at which pressure in the water is zero with respect to atmospheric pressure. Also termed groundwater surface, groundwater elevation, free water elevation, phreatic surface, and water table.

Geohydrology: A term, often used interchangeably with hydrogeology, referring to the hydrologic or flow characteristics of subsurface waters (*see* Hydrogeology).

Ghyben-Herzberg Principle: A principle that accounts for the existence of a body of fresh water floating on sea water within an aquifer because of the different densities. Generally speaking, freshwater extends to a depth about 40 times the height that the fresh water table is found above sea level. Conversely, lowering the freshwater table by 30.5 cm (1 ft) causes seawater to rise 13 m (42 ft) within the aquifer.

Gravel-Packed Well: A well in which coarse-grained filter material (such as sand or gravel) is placed in the annular space between the casing and the borehole to increase the effective diameter of the well and to prevent fine-grained material from entering the well during pumping.

Groundwater: (1) That part of the subsurface water that is in the saturated zone, or (2) loosely, all subsurface water as distinct from surface water. Also spelled as two words, ground water.

Groundwater Barrier: A body of material that is impermeable, or has a relatively low permeability, and that occurs below the land surface in such a position that it impedes the horizontal movement of groundwater and consequently causes a pronounced difference in the level of the water table on opposite sides of it.

Groundwater Basin: An area underlain by permeable materials that are capable of storing or furnishing a significant water supply; the basin includes both the surface and the permeable materials beneath it.

Groundwater Budget: A numerical account of the recharge, discharge, and storage of an aquifer, part of an aquifer, or system of aquifers; the groundwater equation.

Groundwater Divide: A ridge in the water table or other potentiometric surface from which groundwater moves away in both directions.

Groundwater Elevation: The elevation at which pressure in the water is zero with respect to atmospheric pressure. Also termed groundwater surface, free water elevation, free water surface, phreatic surface, and water table.

Groundwater Equation: The equation that balances the groundwater budget.

Groundwater Mound: A rounded, mound-shaped surface in a water table or other potentiometric surface that builds up as a result of the downward percolation of water.

Groundwater Storage: (1) The quantity of water in the saturated zone, or (2) water available only from storage as opposed to capture.

Groundwater Surface: The elevation at which pressure in the water is zero with respect to atmospheric pressure. Also termed groundwater elevation, free water elevation, free water surface, phreatic surface, and water table.

Groundwater Trough. A depression in the water table or other potentiometric surface caused by a flow of groundwater into a stream, drainage ditch, or a line of pumping wells.

Head: (1) The sum of the elevation head, pressure head, and velocity head at a given point in an aquifer; or (2) the water level elevation in a well, or elevation to which water in a flowing well rises in a pipe extended high enough to stop the flow. Expressed in units of length. Also called hydraulic head.

Hydraulic Conductivity: A measure of ease with which a fluid can move through pore spaces and fractures. Is determined for a unit volume of fluid water that will move in unit time under a unit hydraulic gradient through a unit area measured perpendicular to the direction of flow. Is reported in units of length per time. Sometimes called coefficient of permeability or permeability.

Hydraulic Gradient: Slope of the water or potentiometric surface. The change in static head per unit of distance in a given direction. If not specified, the direction generally is understood to be that of the maximum rate of decrease in head.

Hydraulic Head: *See* Head.

Hydrogeology: The science that deals with subsurface waters and related geologic aspects of surface waters. Also used in the more restricted sense of groundwater geology only (*see* Geohydrology).

Hydrograph: A graph showing stage, flow, velocity, or other characteristics of water with respect to time. A stream hydrograph commonly shows rate of flow; a groundwater hydrograph shows water level or head.

Hydrologic Budget: An accounting of the inflow to, outflow from, and storage in a hydrologic unit, such as a drainage basin, aquifer, soil zone, lake, or reservoir; the relationship among evaporation, precipitation, runoff, and the change in water storage, expressed by the hydrologic equation.

Hydrology: The science that relates to the distribution and circulation of naturally occurring water on and under the Earth's surface.

Impermeable: A condition of geologic material that renders it incapable of transmitting significant quantities of water under the pressure differences ordinarily found in subsurface water.

Incidental Recharge: Recharge that occurs from facilities not designed for groundwater recharge (such as irrigation and septic tanks) and by changes in vegetation for purposes other than increasing recharge.

Infiltration: The flow or movement of water through the soil surface into the ground.

Infiltration Capacity: The maximum or limiting infiltration rate.

Infiltration Rate: The rate at which a soil under specified conditions absorbs falling rain, melting snow, or surface water, expressed in depth of water per unit time (cm/s, in./h, etc.).

Injection Well: A well used for introducing water into an aquifer. Technique used to stop seawater intrusion, replenish an aquifer, or dispose of cooling or wastewater.

Interstice: A pore or open space in rock or granular material, not occupied by solid matter. It may be occupied by air, water, or other gaseous or liquid material. Also called void, or void space.

Intrinsic Permeability: A measure of the relative ease with which a porous medium can transmit a liquid under a potential gradient. It is a property of the medium alone and independent of the nature of the liquid and of the force field causing movement. It is a property of the medium that depends upon the shape and size of the pores.

Leakance: The ratio K'/b', in which K' and b' are the vertical hydraulic conductivity and thickness, respectively, of the confining beds.

Leaky Aquifer: *See* Semiconfined Aquifer.

Managed Aquifer Recharge: The intentional recharge and storage of water into an aquifer for subsequent recovery or for environmental benefits.

Mathematical Model: Mathematical equations, based on conservation of mass and the physics of groundwater flow, that can be used to simulate the response of groundwater flow systems to natural or artificially made hydrologic stresses.

Mining: The process, deliberate or inadvertent, of extracting groundwater from a source at a rate so that the groundwater level declines persistently, threatening actual exhaustion of the supply.

Operational Recovery Efficiency (ORE): The efficiency for a single recharge–pumping cycle as long as the recovered water meets target water quality criteria in terms of total dissolved solids (TDS) or chloride concentration. Also called the recovery efficiency.

Overdraft: The condition of a groundwater flow system where the amount of water withdrawn exceeds the amount of water captured by the system over a period of time. The use of water in excess of the perennial yield.

Oxidation Reduction Potential (ORP): A measure of water's ability to receive or gain electrons and thereby be reduced via chemical reactions. Describes the likelihood that metals will precipitate from groundwater. Describes the ability of water to disinfect. Also known as redox potential.

Perched Groundwater: Unconfined groundwater separated from an underlying body of groundwater by an unsaturated zone. Its water table is a perched water table.

Perched Water Table: The water table of a perched aquifer.

Percolation: The movement, under hydrostatic pressure, of water through the interstices of rock or soil, except the movement through large openings such as caves.

Percolation Rate: The rate, expressed as either velocity or volume per unit of time, at which water percolates through a porous medium.

Perennial Yield: The maximum quantity of water that can be withdrawn annually from a groundwater supply under a given set of conditions without causing an undesirable result.

Permeability: The capacity of a geologic material for transmitting a fluid. The degree of permeability depends upon the size and shape of the openings and the extent of the interconnections.

Permeability Coefficient: The rate of flow of water through a unit cross-sectional area under a unit hydraulic gradient. Also called hydraulic conductivity.

Permeable: A condition of a geologic material that renders it capable of transmitting a significant quantity of water without impairment of its structure.

pH: Hydrogen ion activity; the negative logarithm of the hydrogen ion concentration. A number used to express the acidity of solutions, including water. A pH value lower than 7 indicates an acidic solution, a value of 7 is neutral, and a value higher than 7 indicates an alkaline solution.

Phreatic Surface: The elevation at which pressure in the water is zero with respect to atmospheric pressure. Also termed groundwater elevation, groundwater surface, free water elevation, free water surface, and water table.

Phreatic Water: A term that originally was applied only to water that occurs in the upper part of the saturated zone under water table conditions (synonym of unconfined groundwater), but has come to be applied to all water in the saturated zone of an unconfined aquifer.

Phreatophyte: A plant that habitually obtains its water supply from the saturated zone, either directly or through the capillary fringe.

Piezometer: A well emplaced to measure and record groundwater levels; usually contains no well screen or a very short well screen.

Piezometric Head: The hydraulic head of fluid in a confined aquifer.

Porosity: (1) An index of the void characteristics of a soil or rock material; degree of perviousness, or (2) the ratio usually expressed as a percentage of the volume of the interstices, whether isolated or connected, in a given quantity of material to the total volume of the material.

Potable Water: Water that is safe and palatable for human use. Freshwater in which any concentrations of pathogenic organisms and dissolved toxic constituents have been reduced to safe levels, and which is, or has been treated so as to be, tolerably low in objectionable taste, odor, color, or turbidity and of a temperature suitable for the intended use.

Potentiometric Surface: An imaginary surface representing the elevation and pressure head of groundwater and defined by the level to which water rises in a well or piezometer. The water table is a particular potentiometric surface.

Pressure Head: Hydrostatic pressure expressed as the height of a column of water that the pressure can support, expressed with reference to a specific level, such as land surface.

Pump Efficiency: The ratio of energy converted into useful work to the energy applied to the pump shaft, or the energy difference in the water at the discharge and suction nozzles divided by power input at the pump shaft.

Pumpage: The quantity of water (or other liquid) pumped over a period of time.

Recharge: Replenishment of groundwater by downward infiltration of water from rainfall, streams, and other sources. Natural recharge is that recharge that occurs without assistance or enhancement by humans. Managed aquifer recharge is that recharge that occurs when people deliberately modify the natural recharge pattern to increase recharge.

Recharge Basin: A basin constructed on the ground surface to receive the discharge from streams, storm drains, or other sources for the purpose of replenishing groundwater supply.

Recharge Well: A single-purpose well to inject water into an aquifer for eventual recovery at another location.

Recovery Efficiency: *See* Operational Recovery Efficiency (ORE).

Redevelopment: Attempts to restore the original yield of a well by various methods.

Saline Water: Water containing more than 10,000 mg/L of dissolved solids.

Saltwater Encroachment: Displacement of fresh surface or groundwater by the advance of saltwater because of its greater density, usually in coastal and estuarine areas but also by movement of brine from beneath a playa lake toward wells discharging freshwater. Encroachment occurs when the total head of the saltwater exceeds that of adjacent freshwater.

Saltwater Intrusion: The movement of saltwater into freshwater aquifers (*see* Saltwater Encroachment).

Sand: (1) Geologic material with grain sizes ranging from 0.0625 to 2.000 mm (0.0025 to 0.079 in.), or (2) a sediment composed primarily of these particles.

Saturated: A condition in which the interstices of a material are filled with a liquid, usually water. It applies whether the liquid is under greater than or less than atmospheric pressure, as long as all connected interstices are full.

Saturated Thickness: The thickness of the saturated portion of a soil or rock.

Saturated Zone: That part of a water-bearing soil or rock in which all voids are filled with liquid that is under pressure equal to or greater than atmospheric.

Saturation: (1) A condition reached by a soil or rock, that holds a liquid such that no more liquid can be held within it. The material is then said to be saturated or in a condition of saturation. In aquifers this is the filling of all pore spaces and voids with liquid, or (2) the condition of a liquid when it is taken into solution the maximum possible quantity of a given substance at a given temperature and pressure.

Seepage: (1) The act or process involving the slow movement of water or other fluid through a porous material such as soil, or (2) the amount of fluid that has been involved in seepage.

Semiconfined Aquifer: An aquifer in which the deposits above and below the aquifer are somewhat permeable, allowing leakage from the aquifer into adjacent layers. Also called a leaky aquifer.

Silt: (1) Geologic material with grain sizes ranging from 0.004 mm to 0.062 mm (0.00016 to 0.0025 in.), or (2) a sediment composed primarily of these particles.

Skin Effect: The reduction in permeability adjacent to the screened or open interval of a well, because of compaction of sediments, chemical reactions that lead to precipitation of metals, or invasion of drilling mud.

Soil Aquifer Treatment (SAT): The use of subsurface soil to transform or remove from the recharge water contaminants that may affect groundwater quality.

Soil Moisture: Water or moisture contained in the soil mantle.

Soil Porosity: The percentage of the soil (or rock) volume that is not occupied by solid particles, including all pore space filled with air and water.

Source Water: The water used for recharge.

Specific Capacity: The rate of discharge of water from the well divided by the drawdown of water level within the well.

Specific Conductivity: With reference to the movement of water in soil, a factor expressing the volume of transported water per unit of time in a given area.

Specific Injectivity: The rate of recharge of water to a well divided by the increase in water level within the well.

Specific Permeability: A factor expressing the permeability of a material; it equals a constant multiplied by the square of representative pore diameter.

Specific Retention: The ratio of the volume of water that the rock or soil, after being saturated, retains against the pull of gravity to the volume of the rock or soil.

Specific Storage: The volume of water that a unit volume of aquifer releases from storage from a unit volume of aquifer material under a unit decline in head. Is due to the expansion of water and compression of the aquifer void spaces and grains.

Specific Yield: The ratio of volume of water that saturated rock or soil yields by gravity to the volume of the rock or soil.

Static Head: The height above a standard datum of the surface of a column of water or other liquid that can be supported by the static pressure at a given point. It is the sum of the elevation head and the pressure head; the velocity head is negligible under conditions to which Darcy's law can be applied.

Storage Capacity: The volume of space below the land surface that can be used to store groundwater. Divided into total storage capacity and available storage capacity.

Storage Coefficient: The volume of water an aquifer releases from or takes into storage per unit surface area of the aquifer per unit change in head. Also called storativity.

Storativity: *See* Storage Coefficient.

Sustained Yield: The volume of groundwater that can be extracted annually from a groundwater basin without causing adverse effects.

Target Storage Volume (TSV): The volume of water intended for recovery during a design recovery period. May include a buffer zone volume that is initially formed and then maintained around the TSV.

Total Dissolved Solids (TDS): The quantity of mineral (salts) in solution in water, usually expressed in milligrams per liter or parts per million (ppm).

Total Suspended Solids (TSS). The dry weight of suspended particles that are not dissolved in water that can be trapped by a filter.

Total Storage Capacity: The total volume of space that could be used to store groundwater.

Transmissivity: The rate at which water is transmitted through a unit width of the aquifer under a unit hydraulic gradient, equal to the hydraulic conductivity times the aquifer saturated thickness.

Unconfined Aquifer: A saturated soil or rock in which groundwater is in direct contact with the atmosphere through the open pore spaces of the overlying soil or rock. The upper groundwater surface is called the water table.

Unconfined Groundwater: Unconfined water is found in the saturated zone whenever the upper surface of the zone forms a water table under atmospheric pressure, free to rise and fall with changes in volume of stored water.

Unsaturated Zone: The zone between the land surface and the water table. It includes the capillary fringe and contains pore water that is under pressure less than that of the atmosphere. It is also called the vadose zone.

Usable Storage Capacity: The quantity of groundwater of acceptable quality that can be economically withdrawn from storage.

Vadose Zone: *See* Unsaturated Zone.

Vadose Zone (Dry) Well: A well machine or hand dug without casing and ending above the saturated zone.

Void Space: *See* Interstice.

Water Quality: (1) The fitness of water for use; and (2) the physical, chemical, and biological characteristics of water relative to a particular use.

Water Table: The upper surface of the saturated zone of an unconfined aquifer at which the pore water pressure is equal to atmospheric pressure. The water table is a particular potentiometric surface. Also termed groundwater surface, groundwater elevation, free water elevation, free water surface, and phreatic surface.

APPENDIX B
NOTATIONS AND SYMBOLS

AF	acre-foot; approximately 1,233 m^3 (325,851 gal.)
ASR	aquifer storage recovery
ASTM	American Society for Testing and Materials
cm	centimeter
°C	degrees Celsius
cfs	cubic feet per second
DBP	disinfection by-products
DO	dissolved oxygen
Eh	oxidation reduction potential
EIS	environmental impact statement
gpm	gallons per minute
g	gram
h	head (pressure or hydraulic)
ha	hectare
HAA	haloacetic acid
kW	kilowatt
kg	kilogram
L	liter
L/s	liters per second
m	meter
m/s	meters per second
m^2	square meter
m^3	cubic meter
mg/L	milligrams per liter
ML/day	megaliters per day
mm	millimeter
MGD	million gallons per day
NEPA	National Environmental Policy Act
NPDES	National Pollutant Discharge Elimination System
NTU	nephelometric turbidity units
pH	hydrogen ion activity
ppm	parts per million
Q	discharge (yield)
RO	reverse osmosis
SAT	soil aquifer treatment
TDS	total dissolved solids
THM	trihalomethanes
TOC	total organic carbon
TSS	total suspended solids
μg/L	micrograms per liter

APPENDIX C
MAR REGULATIONS IN THE UNITED STATES

C.1 FEDERAL LAWS AND REGULATIONS RELATED TO MAR

In general, federal laws and regulations related to MAR projects are relatively limited. MAR projects that include recharge wells are subject to federal underground injection control (UIC) regulations. MAR projects that recharge via surface facilities (such as an infiltration basin) are not regulated by the USEPA, although certain permits may be required for the discharge to surface water [National Pollutant Discharge Elimination System (NPDES) permit] and for alterations to streambeds (Section 404 of the Clean Water Act) (Fort 2008).

C.1.1 Federal Regulations for Recharge Wells. The regulations of the federal UIC program prohibit recharge (injection) of water that does not meet primary drinking water standards (Aquifer Storage Recovery Technical Advisory Group 2002). UIC regulations may be administered by the USEPA or by the acting state environmental agency, depending on whether the individual state has "primacy" or primary enforcement authority (Figure C-1). According to UIC regulations, recharge (injection) wells are classified as Class V injection wells, one of five classifications of regulated injection wells (USEPA 2016). The recharge wells are typically authorized by rule if (a) the well owners or operators submit basic well information to the appropriate agency (USEPA or state), and (b) the well recharge does not endanger underground sources of drinking water (USDWs).

However, the regulating agency may require an individual permit that includes operating requirements to ensure protection of USDWs. Additional regulations adopted by primacy states for recharge wells vary by state, as suggested by the discussion on selected state regulations in Section C.2. State-specific recharge well regulations do not supersede federal regulations that prohibit USDW endangerment. As of 2007, nine states require water used for recharge to be potable or treated to national or state standards (USEPA 2016). It should be noted that potable water is defined differently from state to state. In this document, *potable* refers to water of high quality posing no health risk when consumed.

The following additional types of water are allowed to be used for recharge by some primacy states: (a) treated effluent, (b) untreated surface and groundwater, and (c) reclaimed water; all of these types of water are subject to state MAR regulations for recharge or state water criteria. However, state-specific MAR regulations cannot be less stringent than federal regulations. Specifically, 40 CFR 144.12(a) states

> *No owner or operator shall construct, operate, maintain, convert, plug, abandon, or conduct any other injection activity in a manner that allows the movement of fluid containing any contaminant into underground sources of drinking water, if the presence of that contaminant may cause a violation of any primary drinking water regulation under 40 CFR part 142 or may otherwise adversely affect the health of persons.*

The type and quality of injected fluid, called *injectate*, and the geology affect the potential for endangering a USDW. The USEPA (2016) identified the following potential concerns:

(a) Pathogens may enter an aquifer if injectate is not disinfected, especially in those states that allow injection of raw water and treated effluent into the aquifer. The fate of microbes and viruses in an aquifer is relevant to the quality of the ambient groundwater and the recovered water.
(b) Disinfection by-products can form in situ when water is disinfected before injection. To avoid this problem, soluble organic carbon should be removed from the injectate before disinfection. If not, a chlorinated disinfectant may react with the carbon to form contaminating compounds, including trihalomethanes and haloacetic acids.
(c) Chemical differences between the injectate and the receiving aquifer may trigger chemical reaction and increase health risks when arsenic and radionuclides in the geologic matrix interact with injectate that has a high reduction–oxidation potential.
(d) Carbonate precipitation in carbonate aquifers can clog wells when the injectate is not sufficiently acidic.

In other words, preventing changes in the chemical and/or microbial composition of the groundwater is not the goal of the federal UIC program. Rather, the goal is to prevent chemical and/or microbial changes that adversely affect the public health of future users of the groundwater.

Overall, MAR activities have not negatively impacted USDW and, in some cases, have resulted in improved water quality. A study of Class V wells conducted by USEPA in 1999 did not identify cases of USDW contamination by recharge wells used in MAR projects (USEPA 1999). One exception resulted from the USEPA lowering the drinking water standard for arsenic from 50 parts per billion to 10 parts per billion in 2001. As a result of the changed standard, testing at some MAR recharge operations in the 1999 study showed concentrations of arsenic exceeding the new standard. Some tested MAR recharge operations had manganese and iron concentrations exceeding USEPA's National Secondary Drinking Water Regulations. Additionally, the presence of disinfection by-products has occurred in USDWs because of ASR activities. Recharge can have positive impacts on the water quality of USDWs if the injected water has, in some cases (NGWA 2014), improved ambient water quality of aquifers.

C.1.2 SDWA for the Recovered Water. USEPA (2016) notes that the UIC program does not regulate water production even though the MAR recharge systems may include production of the

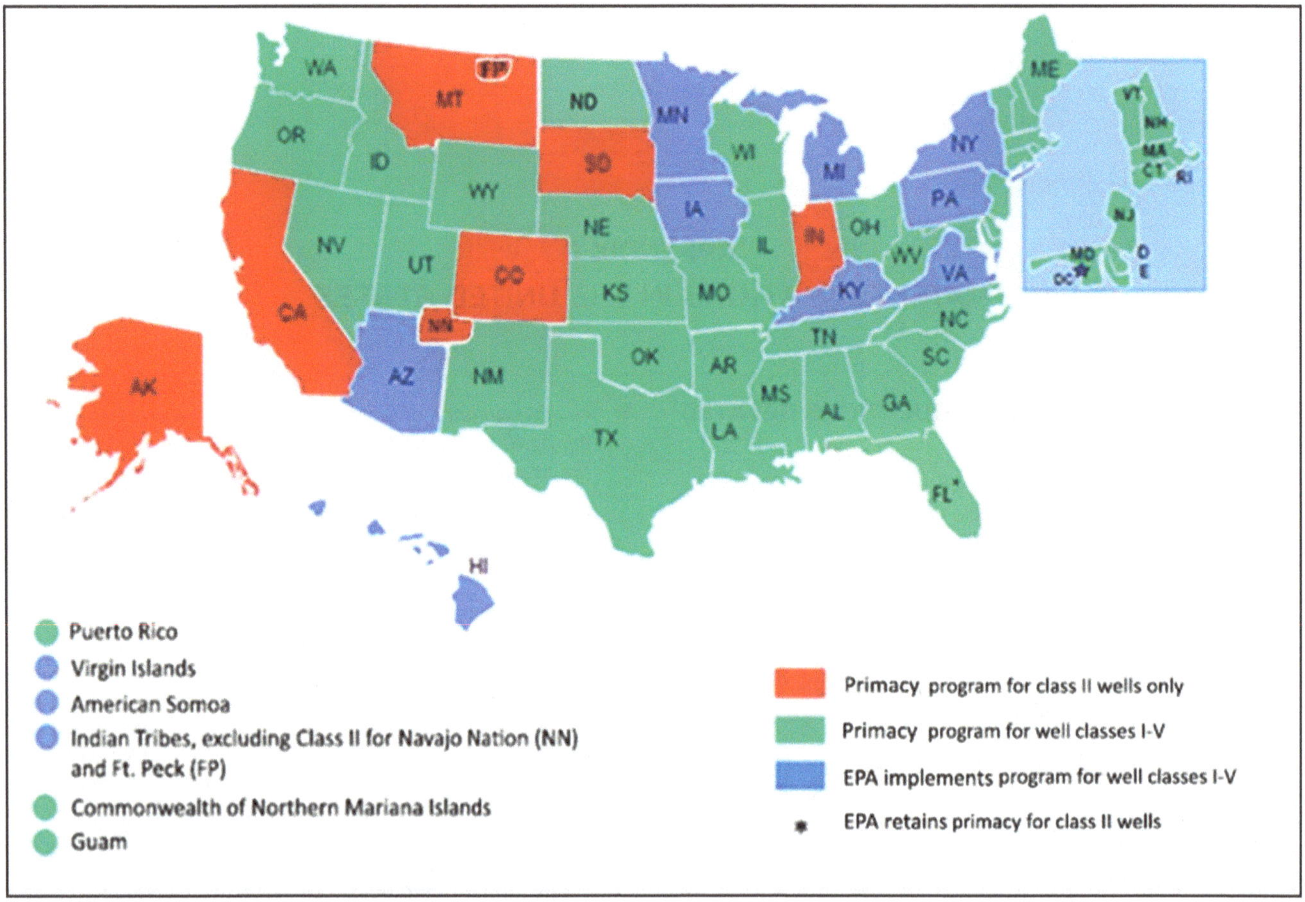

Figure C-1. States, territories, and tribes with UIC primacy.
Source: USEPA. https://www.epa.gov/uic/primary-enforcement-authority-underground-injection-control-program.

injected water. If the recovered water is used for potable supplies, then the federal Safe Drinking Water Act (SDWA) applies to produced water, which requires that water recovered from a MAR dual-purpose well must meet primary drinking water standards before being put into the water distribution system. Currently, the water distributed within a municipal water system must comply with more than 80 established contaminant standards (Aquifer Storage Recovery Technical Advisory Group 2002).

The health-based and enforceable maximum contaminant levels (MCLs) are established by the USEPA for numerous chemicals and microbial pathogens under the authority of the SDWA. Such MCLs established by the SDWA carry risks of adverse health outcomes in a range from 1 in 1 million to 1 in 10,000. All state governments are required to adopt these SDWA standards at levels that are at least as stringent as those specified by the USEPA.

Because most chemical and microbial contaminants are regulated at the entry point to the distribution system, the water recovered from a MAR well must meet these standards before entering the distribution system. If the recovered water does not meet the SDWA standards, the MAR operation must be halted or the recovered water must be subjected to additional treatment before it enters the distribution system. Other chemical and microbial contaminants have potential to increase in concentration at locations in the distribution system; therefore, they are regulated at points in the distribution system. Such contaminants include microorganisms, such as coliform bacteria, and chemicals, such as disinfection by-products and corrosion by-products. Hence, water recovered from the ASR well would need to be of sufficient quality to allow a utility to meet these standards (Aquifer Storage Recovery Technical Advisory Group 2002).

C.2 MAR REGULATIONS IN SELECTED STATES

C.2.1 Arizona Managed Aquifer Recharge Regulations. (modified from Swiezckowski et al. 2007)

In 1980, the Arizona legislature enacted the Arizona Groundwater Management Act (Arizona Revised Statutes §§ 45-401 to 45-704). This work was done to help control the severe groundwater overdraft in certain areas of the state, called active management areas or AMAs, to provide a framework for the allocation of the state's groundwater resources, and to offset the use of groundwater and promote the use of renewable water resources. To accomplish these goals, the legislature created the Arizona Department of Water Resources (ADWR) (ARS §§ 45-101 to 45-116) and, in later legislation, Arizona's recharge program in the form of the Underground Storage and Recovery Act. The recharge program was legislatively modified twice to the existing Underground Water Storage, Savings and Replenishment Program (ARS §§ 45-801.01 to 45-898.01) and passed in 1994. The ADWR administers this program with consultation for water quality issues from the Arizona Department of Environmental Quality (ADEQ) through the ADEQ's Aquifer Protection Permit (APP) program (ARS §§ 49-241 to 49-252). The goals of the recharge program are to promote the use of renewable supplies, to augment the water supply, to help reduce groundwater overdraft, to provide for the efficient use of all water resources by allowing water to be transported by storing water in one location, but recovering a like quantity elsewhere, and to accommodate seasonal demand for water through recharge and recovery.

The ADWR, through the recharge program, imposes administrative and hydrologic requirements on entities applying for and operating an underground storage facility (USF). As part of the USF permit application process, applicants are required to submit extensive technical information to demonstrate that the full permitted volume for the proposed USF is hydrologically feasible and will not cause unreasonable harm to current surrounding land and water users. As part of the application process, the applicant must demonstrate that the proposed USF is sited and will be operated such that recharge is the principle purpose of the facility. Unlike incidental recharge from agricultural irrigation or leakage from artificially constructed bodies of water, the primary purpose of any permitted recharge facility must be to add water to the underlying aquifer. Recharge program policy does allow facilities to have multiple uses, but recharge must always be the primary purpose. Once an application is approved and a USF permit is issued, the permittee is required to schedule a pre-recharge inspection with the department after all associated recharge and monitoring components have been built or installed and before recharge water is delivered to the facility. As soon as the department has given the permittee approval for recharge to commence, the permittee must comply with a series of ongoing operating, monitoring, and reporting requirements, which include submission of quarterly and annual facility reports to the department. After the permittee's reporting has been reviewed for permit compliance, the department issues long-term storage credits for the water that has been stored and is eligible to be recovered.

There are two types of USF permits issued by ADWR: constructed and managed. A *constructed facility* is designed and constructed to store water underground by various methods such as recharge and/or vadose zone wells, trenches, or basins. A *managed facility* is designed to use the natural channel of a stream to store water through the artificial and controlled releases of water other than naturally present surface water. Facilities can be permitted to recharge effluent, surface water, and/or Central Arizona Project (CAP) water, which is the portion of Arizona's allocation of Colorado River water delivered through the CAP Canal. Since the creation of the recharge program in 1986, ADWR has issued 170 USF permits, including permit modifications, throughout the state of Arizona. Currently under the recharge program, there are 59 active USFs, 53 constructed facilities and 6 managed facilities, most of which are located within AMAs. Of the state's permitted constructed recharge facilities, 27 use basins for recharge, 16 use recharge and/or vadose zone wells, 1 facility uses trenches, and the remaining 15 facilities use a combination of these recharge methodologies. The permitted volumes for the active recharge facilities range from 8.3×10^4 m^3 (67 acre-ft) per year to 2.5×10^8 m^3 (20. $\times 10^4$ acre-ft) per year. Of those 59 facilities, 32 are permitted to recharge effluent, 18 are permitted to recharge CAP water, and 9 recharge a combination of source waters. To date, more than 2.5×10^9 m^3 (2.0 million acre-ft) of water have been stored at permitted USFs statewide.

C.2.2 California Groundwater Basin Management and MAR Projects. The California Department of Water Resources (DWR) defines 515 groundwater basins and subbasins in the state. Groundwater use and management of these basins is not centrally regulated, but rather controlled by a series of complex laws and establishment of local agencies. In general, California groundwater can be extracted for beneficial use by any overlying landowner without limitation (subject to a few exceptions, such as an adjudicated basin, described in this section). These overlying landowner rights, the California Water Code, and additional California laws directly impact the permitting and implementation of MAR projects in the state.

Local water districts typically have authority to implement some type of MAR project. The California Water Code enables the formation of numerous types of water districts with varying powers and duties relating to groundwater management, such as California Water Districts, Municipal Water Districts, County Water Districts, Water Replenishment Districts, Water Storage Districts, Reclamation Districts, Irrigation Districts, Water Conservation Districts, and other public and private water districts. In particular, the Water Replenishment District Act (and amendments) established such districts to obtain supplemental water supplies to replenish an overdrafted groundwater basin. In addition, a few special water districts have been established by California law with specific groundwater management powers unique to a certain area or groundwater basin. One such example is the Orange County Water District, operating an advanced groundwater recharge and monitoring system and the nation's largest indirect potable reuse project (involving MAR with recycled water).

There are 22 California groundwater basins in which groundzwater rights have been allocated by the courts. Management of these basins, referred to as adjudicated basins, is typically conducted by a court-appointed Watermaster, who has powers similar to those of a water management district. For adjudicated basins, methods and/or amounts of groundwater basin replenishment for a MAR project may be defined or mandated by the court and Watermaster.

C.2.2.1 MAR Projects as a Component of California Groundwater Management Plans. California MAR projects are often implemented under the umbrella of a Groundwater Management Plan (GWMP) or an Integrated Regional Water Management Plan (IRWMP). GWMPs were authorized by California Water Code Section 10750 in 1993 (later amended in 2002) to allow local public agencies to develop a plan that includes MAR projects, including conjunctive use. In 2002, California provided incentives for local agencies to manage groundwater collaboratively through an Integrated Water Resources Management Plan (IRWMP). The intent of the IRWMP program is to encourage integrated regional water management strategies, including MAR. In the absence of such management plans, MAR projects can also be implemented under coordinated agreements, whereby multiple water purveyors may jointly participate in groundwater management activities.

C.2.2.2 Local Ordinances and Agreements Affecting MAR Projects. Various additional California laws and government agreements (for example, City of Los Angeles v. City of San Fernando, 1975, and the 1994 DWR Monterey Agreements) combine to protect certain rights of a MAR project sponsor. For any party importing nonlocal water for aquifer recharge, that party generally has the right to pump an equivalent amount of groundwater from storage (less losses), irrespective of other pumping in the basin. Several California counties have adopted groundwater ordinances that directly impact MAR projects and may require a permit for MAR implementation. For example, Madera County adopted an ordinance in 2001 regulating any groundwater banking activities in the county that are outside of the jurisdiction of a local water agency or incorporated city.

C.2.2.3 Water Quality Permits in California. Although the state does not regulate the extraction of groundwater, state agencies are responsible for the protection and regulation of groundwater quality. The guiding legislation is the California Clean Water Act, enacted in 1968 and generally referred to as Porter–Cologne or the antidegradation policy. This law gave authority to the State Water Resources Control Board (SWRCB) and nine regional water quality control boards (RWQCBs) to preserve and enhance all beneficial use of surface water and groundwater. The RWQCBs develop basin plans for each of their regions defining beneficial uses and setting water quality objectives. MAR projects are not always consistent with details in some of the basin plans; for example, aquifer recharge is not always designated as a beneficial use of surface water.

The RWQCBs issue waste discharge requirements (WDRs), which function as a permit for any discharge to land or groundwater. Although this program was originally envisioned for wastewater, WDRs or water recycling requirements (WRRs) may be required for a MAR project. In the event that constituents in recharge water do not meet basin plan water quality objectives or are higher than in local groundwater, MAR projects can be impeded. These restrictions can occur even if the recharge water meets all drinking water quality standards. In addition, even more stringent objectives may apply to MAR projects involving direct injection.

Permits issued by the RWQCB may also be required for a MAR project. If aquifer recharge occurs within close proximity to recovery wells, the drinking water system may be permitted by the SWRCB Division of Drinking Water (DDW) as groundwater under the direct influence of surface water (GUISW) regulations and may require additional treatment for the recovered water. If the project involves the recharge of treated wastewater or recycled water, an RWQCB permit is required under the state's Groundwater Recharge Reuse Project (GRRP) regulations. The GRRP must first be reviewed and approved by DDW via the issuance of findings and conditions, which are incorporated into the RWQCB.

C.2.2.4 Additional California State and Local Agency Involvement in MAR Projects. Numerous additional state and local agencies may have requirements applicable to a specific MAR project. Because many MAR projects in California are sited along drainageways and/or involve stormwater, flood control agencies are often involved. A MAR project that involves any alteration to a streambed may require a permit from the California Department of Fish and Game (CDFG). Any wells included in MAR projects require a permit from DWR (and, in some cases, other local agencies) and are also subject to state and local well construction standards.

C.2.2.5 California Environmental Quality Act. For a MAR project to comply with the California Environmental Quality Act (CEQA), an environmental assessment is required. Depending on the nature of the project, this assessment may involve a simple checklist, a mitigated negative declaration, or a detailed environmental impact report. Analyses typically involve potential environmental impacts from construction of project components, such as a pipeline or infiltration basin, and potential long-term impacts from MAR operation such as changes to water quality and quantity.

C.2.2.6 Involvement of Federal Agencies in California MAR Projects. Federal agencies may be involved in a California MAR project, especially if it involves federal lands within the state. This section summarizes some examples of federal involvement. This description is not comprehensive and, depending on the location and details of any particular MAR project, additional agencies may be involved.

If the recharge project involves jurisdictional waters, then any activity such as constructing an infiltration basin may be subject to the US Army Corps of Engineers (USACE) nationwide 404 Permit. The US Fish and Wildlife Service (USFWS) may be involved if the protection of endangered species is an issue on a MAR project. A federal Endangered Species "Take" permit (pursuant to Section 10) or Consultation Permit (pursuant to Section 7) is required for any activity that occurs on US federal lands and involves the destruction or "taking" of an endangered or threatened species (for example, the desert tortoise is a concern over much of California's deserts). The project would also be subject to the requirements of the National Environmental Policy Act (NEPA).

Because USEPA the US Environmental Protection Agency (USEPA) has not delegated full primacy to California for the UIC program, USEPA remains the permitting authority for injection wells (including Class V recharge wells) under the program. A MAR project may involve stormwater drainage wells or aquifer recharge wells that would be classified as Class V recharge wells. Although USEPA does not currently issue Class V permits, any owner of a Class V recharge well must submit information to USEPA and must be included in an inventory.

C.2.2.7 Construction of a MAR Project. The construction of pipelines, basins, access roads or berms, wells, and other project components involves an additional set of permits and regulations that are not discussed herein. Additional agencies that may be involved include the county or city public works department or transportation division, air quality management districts, and others.

C.2.3 Recharge Regulations in Colorado. Colorado has regulations that allow for the underground storage of water in selected alluvial and confined aquifers. The Water Right Determination and Administration Act of 1969 (CCR 37-92-101) allows groundwater recharge and storage, defining aquifer storage as the water that has been placed in the subsurface by other than natural means and with water to which the person placing it has a conditional or decreed right (CCR 37-92-103) (10.7). Recharge activities have been done under augmentation and replacement plans, a mechanism that allows junior appropriators to divert water out of priority. The recharged water creates return flows to hydraulically connected streams and offsets depletions so as to not cause injury to vested senior rights under the state's prior appropriation water rights system. There has been a long history of recharging aquifers in Colorado through augmentation plans, with more than 2.5×10^9 m^3 (2 million acre-ft) of water added to the alluvial aquifer system in the South Platte River Basin in northeast Colorado since 1979 (Topper et al 2004). Most recharge water occurs through leakage from irrigation ditches and unlined reservoirs and ponds. The state Division of Water Resources has developed protocols for recharge structures in the South Platte Basin; these protocols include ditches and ponds whose use will result in accretions to a live stream. Recharge flows must be computed on a daily basis and must include evaporative and consumptive use losses.

In 1995 the state's Division of Water Resources established Artificial Recharge Extraction Rules for the Denver Basin (2 CCR 402-11) to extend the life of the Denver Basin aquifers. The Denver Basin is a large region in the east-central part of the state that includes four sedimentary bedrock aquifers. The 1995 rules apply to the evaluation and processing of applications for permits to extract water that has been recharged into one or more of the Denver Basin aquifers located outside the boundaries of any designated groundwater basin. The 1995 rules establish minimum standards for documenting the nature and amount of recharged water available for extraction and for the permitting and construction of extraction wells. Wells used for recharge and/or extraction must be equipped with suitable totalizing flowmeters. The rules allow for extraction to be from a well up to 8 km (5 mi) from the recharge location in a confined aquifer and up to 0.3 km (1,000 ft) from the recharge location in an unconfined aquifer, provided that the extraction well is within the same contiguous parcel of land. Recharged water can be retained in the aquifer indefinitely or can be removed in an amount up to five times the amount injected in any calendar year, provided that the total amount extracted does not exceed the total amount of water injected. Annual reports must be submitted to the state engineer that describe the timing, location, amounts of water injected and extracted on a weekly basis, and the water level in the well measured on an annual basis.

C.2.4 MAR Rules and Regulations in Nebraska. Water rights are regulated by the Nebraska Department of Natural Resources (DNR). Generally, separate permits are required to recharge water, to store the water, and to recover the water. The type of permits required for recharge varies according to the method of recharge (i.e., surface infiltration or direct injection). Rules and regulations for Incidental and Intentional Underground Water Storage are defined in Title 457, Chapter 16, of the Nebraska Administrative Code. Rules and regulations for Induced Groundwater Recharge are defined in Title 457, Chapter 22, of the Nebraska Administrative Code. If water is to be recharged by recharge (injection) well, a well permit would be required. The DNR defines an injection well as "a well used for injecting water into the underground water reservoir. Not included are wells used for inserting media to repressurize oil or natural gas bearing formations regulated by the Nebraska Oil and Gas Conservation Commission."

C.2.5 MAR Rules and Regulations in New Mexico. Enabling legislation that allows for underground storage and recovery (USR) was passed by the New Mexico legislature in 1999. The Groundwater Storage and Recovery Act (NMSA 1978, §72-5A-2) provides the legal mechanism for public entities to retain rights to withdraw water that is recharged to an aquifer. The legislature specifically found that the "conjunctive use and administration of both surface and groundwaters are essential to the effective and efficient use of the state's limited water supplies" and that groundwater recharge, storage, and recovery have the potential to reduce the rate of aquifer decline, promote conservation, serve public welfare, and lead to more effective use of water resources. Water stored pursuant to the Act is exempt from forfeiture (NMSA 1978, §72-5A-8). Water can be stored pursuant to this statute only by permitted public entities, and several criteria must be met before a permit will be issued (NMSA 1978, §72-5A-6).

Two types of permits are required for enhanced recharge projects: a USR permit from the Office of the State Engineer (OSE) and a groundwater discharge permit from the New Mexico Environment Department (NMED).

Additional permits may be required, depending on the nature and location of the enhanced recharge project. Additional permitting may include water rights transfers, modifications to public water supply systems, approval from flood control agencies, and surface water quality protection.

C.2.5.1 Establishing Water Rights. USR must be conducted in conformance with the requirements of the OSE Underground Storage and Recovery Regulations (19.25.8 NMAC, effective January 31, 2001). More specifically, the law requires a permit issued by the OSE for any governmental agency to construct or operate a recharge, storage, and recovery project (19.25.8.13 NMAC). These regulations govern the application process; the hydrologic, technical, and financial capability requirements; and the permit terms and conditions authorized under the Act. Based on the legal requirements set forth in 19.25.8 NMAC, USR methods are a viable option for conjunctive management of groundwater and surface water resources; however, additional hydrogeologic investigations are necessary to establish water rights accounting and the recoverability of stored water. The OSE decides whether recharged water is fully recoverable or whether an unrecoverable loss occurs.

To obtain a permit, the OSE requires a pilot/demonstration project and a complete capability report for the pilot/demonstration project. The applicant must submit to the OSE, during a preapplication meeting, a complete project proposal that addresses many of the hydrogeologic and technical aspects of the proposed project, including

- Detailed description of the purpose and scope of the proposed project (both the pilot/demonstration project and the full-scale project), the area of hydrogeologic effect, and the method of recharge;
- An inventory and topographic map of existing natural and artificially constructed water bodies, past and present land uses, possible hazardous waste and potential sources of contamination, and possible water rights impairment that may result from the proposed project;
- Complete characterization of the source water, ambient groundwater, and the hydrogeology of the vadose zone and aquifer;
- Scope of work for
  - Monitoring the vadose zone and saturated aquifer, surface water quality, and subsidence;
  - Infiltration testing;
  - Obtaining aquifer parameters; and
  - Groundwater/surface-water modeling;
- Discussion of the methods used to calculate the amount of recoverable water in storage and the method for recovering recharged water; and
- Demonstration that the applicant has a valid water right and meets all relevant state and federal rules and regulations.

After the initial application has been filed and the project proposal has been submitted during the preapplication meeting with representatives from the OSE, a capability report for the pilot/demonstration project must be prepared and submitted to the OSE. The capability report is a comprehensive document that describes the hydrologic, technical, and financial capability of the applicant, and because this permit review process has not been completed previously, it is expected that the report will require considerable time for OSE review. As discussed in 19.25.8.14 NMAC, the data required for the pilot/demonstration project depend on the complexity, size, and timescale of the project.

After the OSE has conducted an initial review and accepted the application for filing, the USR permit applicant must publish a public notice. Any protesters of the application can request that a hearing be held to consider the application and any impairment that may result. A public hearing, if required, is held in accordance with the OSE hearing procedures of 19.25.2 NMAC.

C.2.5.2 Obtaining Discharge Permits. USR projects are subject to approval of a groundwater discharge permit by the NMED Groundwater Quality Bureau unless NMED determines that a groundwater discharge permit is not required. These regulatory requirements are administered by the NMED under the Water Quality Act (NMSA 1978, §74-6-1 et seq.). Discharge permit approvals are issued under the New Mexico Water Quality Control Commission (WQCC) Regulations (20.6.2 NMAC), which cover both groundwater and surface water protection.

The first step in WQCC compliance is filing a notice of intent (NOI) to discharge with NMED (20.6.2.1201 NMAC) and establish Ground Water Quality Bureau and Surface Water Quality Bureau requirements. If the water source contains contaminants that may affect groundwater, as determined by NMED, an approved groundwater discharge permit is required. If a USR project is planned that could affect surface water quality, such as enhanced recharge in a natural watercourse, the Surface Water Quality Bureau could be involved, under requirements for a National Pollutant Discharge Elimination System (NPDES) permit. However, most USR projects involving injection or other recharge to an aquifer only need to address groundwater protection requirements.

Presumably, a range of regulatory compliance requirements and scrutiny are applied to various types of USR projects with different potentials for groundwater quality impacts. For example, groundwater recharge using treated wastewater effluent is likely to be closely examined to ensure that treatment levels are adequate to protect groundwater quality. A discharge permit for treated effluent requires periodic monitoring to ensure that consistent, high-quality treatment is achieved. Recharge of surface water may require monitoring and possible treatment, depending on the water quality of the surface water source. NMED may determine that some USR projects do not have a potential for contaminant impacts, in which case no groundwater discharge permit would be required. Such projects could involve recharge of treated drinking water or increasing recharge of stormwater by channel alterations in a watercourse where natural recharge already occurs.

To obtain a groundwater discharge permit, the WQCC regulations require a complete application, including

- Quantity and quality of the water being recharged and plans for water quality monitoring;
- Locations of watercourses, supply wells, and existing or proposed monitor wells for the project;
- Hydrogeologic description of the aquifer(s) underlying the site, including their groundwater quality;
- A demonstration that the discharge permit will not result in groundwater concentrations that exceed WQCC standards or the presence of any listed toxic pollutant (20.6.2.7.VV NMAC) and a contingency plan to cope with possible system failure;
- For USR using injections wells, a demonstration that the discharge permit meets the WQCC Underground Injection Control requirements in 20.6.2.5000 NMAC for Class V injection wells; and
- A closure plan ensuring continued compliance with groundwater quality standards, including post-closure monitoring and abatement, if needed.

Both the permit applicant and NMED must issue public notice of discharge plan applications, with specific notice provided to adjacent property owners, government entities, and the general public. The public has an opportunity to submit comments and request that a public hearing be conducted. A public hearing is held if NMED determines that there is significant public interest.

The scope of NMED's discharge plan review is established by the policy development work group and during future review of USR applications. Although the WQCC regulations contain exemptions from discharge plan requirements for water that meets numerical standards for water quality (20.6.2.3105.A NMAC), the regulations also grant broad authority to prevent any discharge that may result in hazard to public health [20.6.2.3109.H(3) NMAC]. NMED may have concerns over the possible presence of trace level chemicals in recharge source water, even for source water that meets all drinking water standards. Concerns over unregulated trace constituents are probably most significant for influent sources from treated wastewater or stormwater from developed areas. Although recharge of groundwater from river seepage and irrigation canals is generally considered a benefit to the aquifer, river water may still contain some of these unregulated trace constituents. Applicants for proposed USR projects using either recharge wells or surface infiltration methods need to work with NMED to establish the appropriate discharge plan requirements based on source water quality, recharge methods, and groundwater characteristics.

C.2.6 MAR Rules and Regulations in Texas. In 2015, the Texas legislature enacted H.B. 655 (ASR Bill), which substantially changed the regulatory process for ASR projects in the state Water Code. It defines an aquifer storage and recovery project as a project involving the recharge of water into a geologic formation for the purpose of subsequent recovery and beneficial use by the project operator.

Some of changes are summarized here:

- The pilot project precondition is no longer required for permission;
- The Texas Commission on Environmental Quality (TCEQ) was granted regulatory authority over recharge and recovery of stored water under Class V injection wells;
- An impact assessment of ASR recharge in the receiving aquifer is required from the project operator;
- TCEQ is required to consider if the injected water complies with federal Safe Drinking Water Act (42 U.S.C. Section 300f et seq.) standards, how much of the injected water will be recovered for beneficial use, whether the injected water could degrade the quality of the native groundwater so that it might be harmful or require an unreasonably high level of treatment to be suitable for beneficial use, and the boundary of land where a single aquifer storage and recovery project will be located;
- TCEQ was prohibited from adopting or enforcing groundwater quality protection standards for injected water that are more stringent than applicable federal standards;
- Groundwater conservation districts have limited authority to register recharge wells and can regulate well pumping only if more water is extracted than was injected; and
- Special water districts, such as the Harris–Galveston Subsidence District, Fort Bend Subsidence District, Barton Springs–Edwards Aquifer Conservation District, and the Corpus Christi Aquifer Storage and Recovery Conservation District, were granted some regulatory jurisdiction over ASR projects.

APPENDIX D
CASE STUDIES

Table D-1. MAR Case Study Summary.

	Case study	Year Constructed	Location	Recharge Method	Size
			Surface Spreading Facilities		
1	Bear Canyon Recharge Project	2009	Albuquerque, NM	Surface recharge in natural arroyo	13,640 m^3/day (3 MGD)
2	OCWD Surface Water Recharge Operations	1936 - 2012	Anaheim, CA	Santa Ana River Channel	90,900-163,700 m^3/day (20–36 MGD)
3	Rancho California Water District	2000	Temecula, CA	Off-channel spreading basins	3.1×10^7 m^3/yr (25,000 AF/yr)
4	Tamarack Ranch Recharge Project	2007	Sterling, CO	Surface recharge basins	1.52×10^7 m^3/yr (12,300 AF/yr)
5	Southern Avra Valley Storage and Recovery Project	2008	Tucson, AZ	Surface recharge basins	8.03×10^7 m^3/yr (65,100 AF/yr)
6	Montebello Forebay Spreading Grounds	1937–1955	Pico Rivera, CA (Los Angeles County)	In-stream check dams and off-channel spreading basins	1.64×10^8 m^3/yr (133,000 AF/yr) (avg. 1959 to 2016)
			Subsurface Recharge Facilities		
7	Denver Basin Aquifer Recharge Demonstration Project	1996	Denver, CO	Well (474.3 m; 1,556 ft)	1,582 m^3/yr (1.283 AF/yr) in 4 years
8	OCWD Talbert Gap Seawater Intrusion Barrier	1975	Anaheim, CA	36 wells (depths 30-210m;100–700 ft)	90,900-163,700 m^3/day (20–36 MGD)
9	Phoenix ASR Well	2010	Phoenix, AZ	Well (433 m; 1,420 ft)	11,370 m^3/day (2.5 MGD)
10	Rio Rancho Direct Injection Demonstration Project	2013	Rio Rancho, NM	Well (520 m; 1,700 ft)	4,540 m^3/day (1 MGD)
11	City of El Paso Clean Water Recharge	1984	El Paso, TX	10 Wells (193-269 m; 632–881 ft)	34,100 m^3/day (7.5 MGD)
12	City of Beaverton ASR Program	2001, 2005	Beaverton, OR	3 wells	6,360 m^3/day (1.4 MGD)

CASE STUDY 1 DESCRIPTION

Project Name: Bear Canyon Recharge Project
Project Location (City/State): Albuquerque, New Mexico
Operating Agency: Albuquerque Bernalillo County Water Utility Authority
Project Contact: Katherine Yuhas
Phone: 505-289-3056, Email: kyuhas@abcwua.org

MAR Project Type: ☒ Surface Basin ☐ Subsurface ASR ☐ Vadose Zone Well
This is a surface infiltration project, performed using a natural arroyo. The channel area used for recharge is approximately 156,000 m^2 (168,000 ft^2, or 4 ac.).
Basin Area (acres): NA
Well Depth (ft below ground surface): NA

Flow Rate [million gal./day (MGD)]: 3
Year Completed: 2009
Currently Operational: ☒ Yes ☐ No
Cost ($): $1.4 million Demonstration Phase? Yes ☒ / No ☐

Project Description. The Albuquerque Bernalillo County Water Utility Authority (water authority) is implementing recharge projects for conjunctive management of surface water and groundwater resources using treated San Juan–Chama water imported from the San Juan River Basin and diverted from the Rio Grande. The purpose is to recharge the Santa Fe Group aquifer system of the Middle Rio Grande Basin, establishing a long-term drought reserve.

The New Mexico Office of the State Engineer (OSE) requires that each underground storage and recovery (USR) project be tested first as a demonstration project. If successful, the applicant may apply for a full-scale permit. DBS&A, an environmental services company, has assisted the water authority to successfully obtain USR permits from the OSE, including demonstration and full-scale permits for Bear Canyon Recharge (USR-2).

The Bear Canyon Recharge project involves releasing bank-filtered surface water into an arroyo channel through the 150 m (500 ft) thick vadose zone profile during the winter months (Figure D-1). The permit allows the water authority to recharge up to 3.7×10^6 m^3/yr (3,000 AF/yr). DBS&A designed and implemented the demonstration project, including significant monitoring to demonstrate the effectiveness of the recharge methods. DBS&A installed a transect of heat dissipation sensors and water quality suction lysimeters, three monitoring wells used to monitor water levels and water quality, and neutron logging access tubes. Because of the success of the demonstration project, the OSE established a storage account for the recharged water. This project became the first full-scale permitted recharge project in the state of New Mexico.

The first full-scale recharge event ran from November 2014 through March 2015, and a total volume of 642,150 m^3 (520.6 AF) was recharged. The OSE credited the water authority 622,900 m^3 (505.0 AF) for this recharge event, or 97% of the total volume recharged. The project was operated October 2016 through February 2017, and a total volume of 682,600 m^3 (553.40 AF) was recharged. The OSE credited the water authority 662,130 m^3 (536.80 AF) for this recharge event. The project will be operated in alternate years going forward to minimize operational (monitoring and reporting) costs, with each event including the recharge for two calendar years.

Particular Challenges Encountered or Overcome. Navigating the regulatory requirements was the most challenging aspect of this project, as the Bear Canyon Recharge Project was one of the first recharge projects in New Mexico. DBS&A developed a monitoring program to track the movement of the water through the vadose zone to prove that the recharge methods were successful.

The Bear Canyon Recharge Project was the first project to receive a full-scale permit in the state of New Mexico, and the success of the ongoing project is of benefit to the state's underground storage and recovery program, the water authority, and other entities that are now pursuing these projects in New Mexico.

Details of this project are provided in Ewing et al. (2016, 2015, 2014), Wolf et al. (2016), Moore et al. (2007), and Moore and Stomp (2009).

Acknowledgments. John M. Stomp, III, P.E., Albuquerque Bernalillo County Water Utility Authority

Katherine Yuhas, Albuquerque Bernalillo County Water Utility Authority

Amy Ewing, P.G., Daniel B. Stephens & Associates, Inc.

Chad Johannesen, P.G., Daniel B. Stephens & Associates, Inc.

Chris Wolf, P.G., Daniel B. Stephens & Associates, Inc.

Michael J. Bitner, formerly with Daniel B. Stephens & Associates, Inc.

Stephanie Moore, Daniel B. Stephens & Associates, Inc.

John Kay, P.G., formerly with Daniel B. Stephens & Associates, Inc.

CASE STUDY 2 DESCRIPTION

Project Name: Staffing of OCWD Surface Water Recharge Operations
Project Location (City/State): Anaheim, California
Operating Agency: Orange County Water District
Project Contacts: Adam Hutchinson and Don Houlihan
Phone: 714-378-3214 (AH), 714-378-3355 (DH),
Email: ahutchinson@ocwd.com, dhoulihan@ocwd.com
MAR Project Type: ☒ Surface Basin ☐ Subsurface ☐ Vadose Zone Well
Basin Area (acres): 1,000
Well Depths (ft below ground surface): NA

Figure D-1. Water in the Bear Canyon Recharge Project.

Flow Rate (MGD): 20–36
Year Completed: 1936–2012
Currently Operational: ☒ Yes ☐ No
Cost ($): N/A
Demonstration Phase? Yes ☐ / No ☒

Project Description. The Orange County Water District began developing its surface water recharge system in 1936, when it purchased its first parcel of the Santa Ana River Channel. Over the years, this system has expanded and now covers more than 610 hectares (1,500 ac) (Figure D-2). Along with land purchases, the district invested in infrastructure to maximize the ability of the facilities to recharge water, including four rubber dams, more than 10 km (6 mi) of pipelines, eight pump stations, and a fleet of earthmoving equipment.

The number and type of staff required to run this operation have evolved over the years. In the early years, the focus was on heavy equipment and landscaping operations. As the system infrastructure became more complex, additional staff, with water system operations and maintenance experience, were added. The current number of full-time staff is 21, with the following breakdown:

- Director of Field Operations 1
- Senior Administrative Assistant 1
- Scheduler/Planner 1
- Recharge Operations Supervisor 1
- Recharge Operator 1
- Assistant Recharge Operator 2
- Maintenance Supervisor 1
- Maintenance Technician Grade 2 2
- Maintenance Technician Grade 1 1
- Senior Heavy Equipment Mechanic 1
- Apprentice Heavy Equipment Mechanic 1
- Operations Supervisor 1
- Senior Heavy Equipment Operator 3
- Heavy Equipment Operator 3
- Landscape Supervisor 1

Particular Challenges Encountered or Overcome. Finding skilled staff for many of the positions required is relatively easy, except for recharge system operators. Recharge system operators need a unique set of skills that includes measuring surface water flows, routing water through open channels and pipelines, filling and draining basins, operating pumps, operating the system via computer, and recording and analyzing the flow and recharge data.

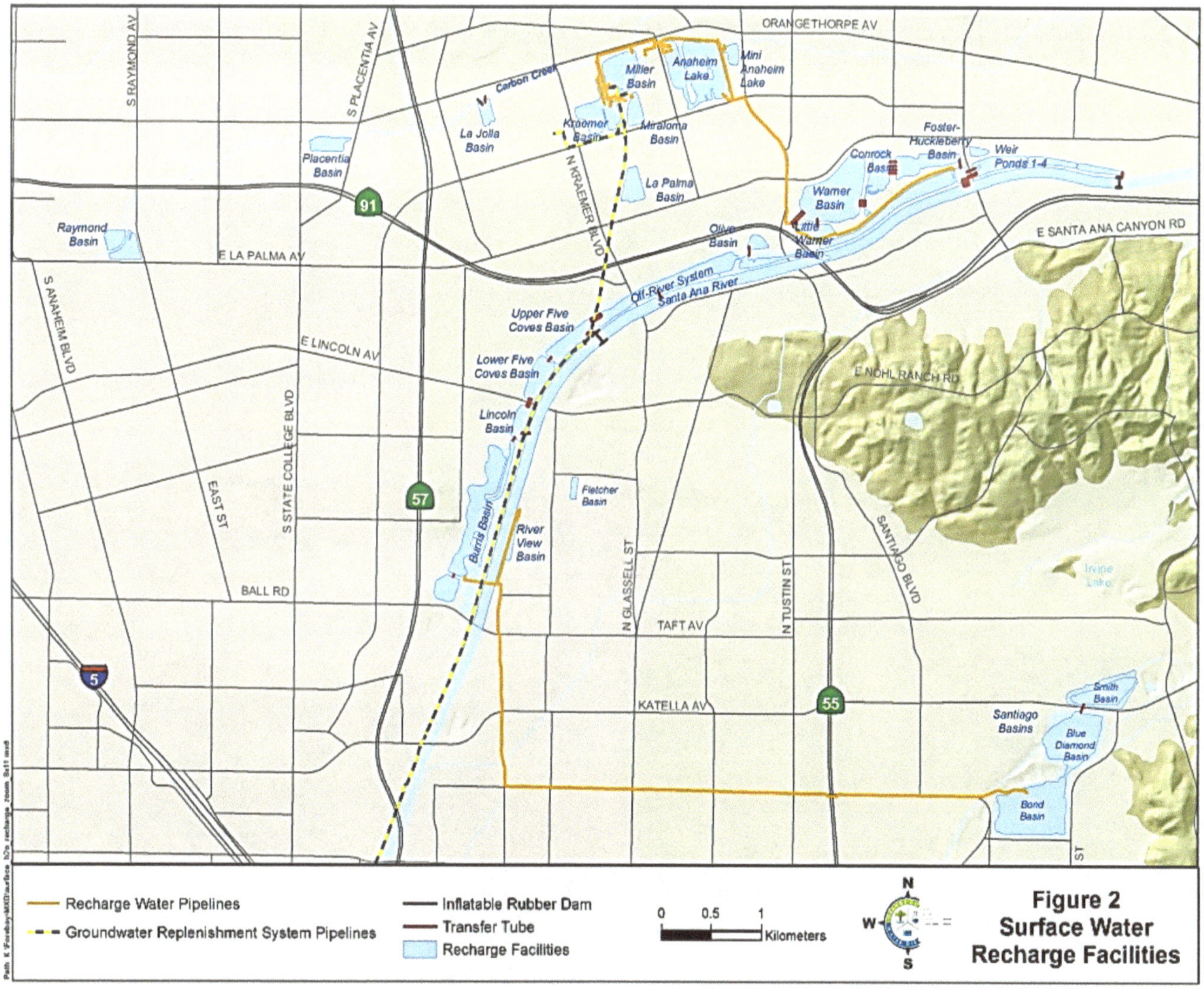

Figure D-2. OCWD surface water recharge facilities.
Source: Orange County Water District.

CASE STUDY 3 DESCRIPTION

Project Name: Rancho California Water District
Project Location (City/State): Temecula, California
Operating Agency: Orange County Water District
Project Contact: Phyllis Stanin
Phone: NA
MAR Project Type: ☒ Surface Basin ☐ Subsurface
☐ Vadose Zone Well
Basin Area (acres): 113 well depths (ft bgs): NA
Flow Rate (AF/year): 25,000
Year Completed: 2000
Currently Operational: ☒ Yes ☐ No
Cost ($): NA
Demonstration Phase? Yes ☐ / No ☒

Project Description. This project's goal is to optimize recharge and recovery in basins near Temecula, California balancing short-term recovery with long-term storage.

Rancho California Water District operates five groundwater basins for MAR, referred to as the Upper Valle de los Caballos Recharge and Recovery Facility (VDCR/RF), to provide a sustainable groundwater supply for the greater Temecula and Murrieta area of California. The Upper VDCR/RF covers 45.2 hectares (113 ac) in the easternmost area of the Pauba Valley near Vail Lake for recharge of raw (untreated) imported water deliveries from Metropolitan Water District of Southern California in addition to Vail Lake release to recharge the large groundwater basin underlying the Pauba Valley. Figure D-3 shows water being released into one of the ponds.

Recharge occurs primarily during the winter months for subsequent recovery during peak demand periods. The Upper VDCR/RF is operated with numerous shallow production wells sited adjacent to and downgradient from the five recharge basins. The retention time from recharge to recovery wells allows for soil aquifer treatment of the untreated surface water. Recharged water that travels in the groundwater system beyond the VDCR/RF remains in the aquifer for longer term storage and is recovered by both shallow and deep production wells located throughout the valley.

The district developed and applied a local-scale five-layer numerical groundwater flow model (the MODFLOW grid, shown in Figure D-4) to simulate aquifer response to historical recharge events to better track recharged water. Numerous operational scenarios were evaluated, including increased recharge in certain ponds or portions of ponds, use of wells adjacent to the basins to control groundwater mounding, and selective operation of key downgradient wells to increase recovery of the recharge water. Particle tracking was used to determine the fate of the recharged water.

The Rancho California Water District worked with the California Department of Public Health (CDPH) to manage recharge and recovery locations and timing to target recharge for specific purposes. The improved understanding also allowed the district to avoid rejected recharge into local drainages and to achieve higher volumes of recharge overall.

Figure D-3. Rancho California Water District recharge basin.

CASE STUDY 4 DESCRIPTION

Project Name: Tamarack Ranch Recharge Project
Project Location (City/State): Between Sterling and Julesburg, Colorado
Operating Agencies: Colorado Division of Wildlife, Colorado Water Conservation Board, Northern Colorado Water Conservancy District, Colorado State University, and Colorado Division of Water Resources
Project Contact: Jon Altenhofen, Northern Water
Phone: 970-622-2236, Email: jaltenhofen@ncwcd.org
MAR Project Type: ⊠ Extraction (retiming) wells next to the river with surface recharge basins
Pumping Wells: 16 wells yielding 5,700–11,000 L/min (1,500–3,000 gpm) each
Well Depths (ft below ground surface): 80–120
Basins: 11 totaling 80 ac
Flow Rate: 50 cfs (10,000–20,000 AF/yr)
Year Completed: 2007
Currently Operational: ⊠ Yes ☐ No
Cost ($): Approx. $3.5 million to build
Demonstration Phase? Yes ☐ / No ⊠

Project Description. Streamflow augmentation through MAR and retiming has been conducted at the Tamarack Ranch State Wildlife Area in Colorado since 1997. The site is located 48.3 km (30 mi) from the Colorado and Nebraska state line (Figure D-5). The Tamarack I and II MAR projects include at Tamarack Ranch 11 recharge ponds that are fed by 16 wells (Figures D-6 and D-7). In addition to the river augmentation for Platte River Recovery Implementation Program (PRRIP) credit, multiple on-site waterfowl habitat benefits have resulted.

The PRRIP is a tristate cooperative agreement among Colorado, Wyoming, Nebraska, and the US Department of Interior focused on maintaining habitats along the central Platte in Nebraska that benefit endangered bird and fish species (piping plover, pallid sturgeon, whooping crane, and interior least tern). Annual accretion credits benefiting the species from the Tamarack I project between 2007 and 2015 have averaged 9,694,000 m^3 (7,859 AF) (Colorado's Annual Report for 2015).

The Tamarack II project is designed to replace depletions caused by population growth impacts since 1997. Between 1997 and 2015, Colorado's population has increased 2.2% per year, which translates to an additional depletion to the river of 2,429,000 m^3 (1,969 AF) average per year. Under Tamarack II, additional accretion credits between 2007 and 2015 have averaged 5,461,000 m^3 (4,427 AF) per year, more than replacing the depletions. Therefore, the net total summer augmentation from MAR for both Tamarack projects has averaged 15,200,000 m^3 (12,300 AF) per year.

Colorado's contribution to the PRRIP 13-year program (2007–2019) is about $28 million, which includes administrative, Nebraska land issues, adaptive management research along with Tamarack construction, on-site personnel, and water rights costs. Colorado water users also paid for construction and are paying all

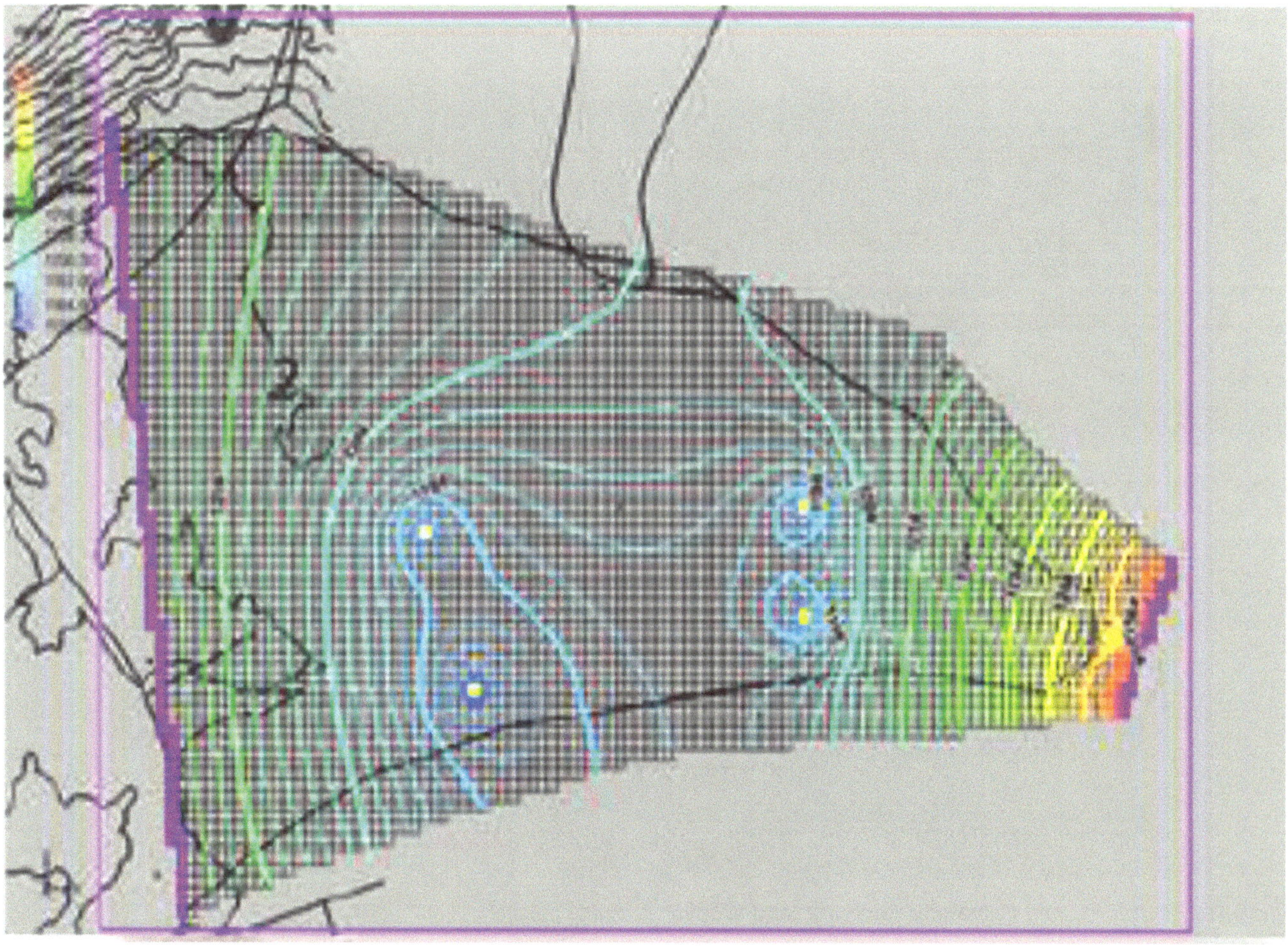

Figure D-4. MODFLOW model grid of Rancho California Water District recharge project.

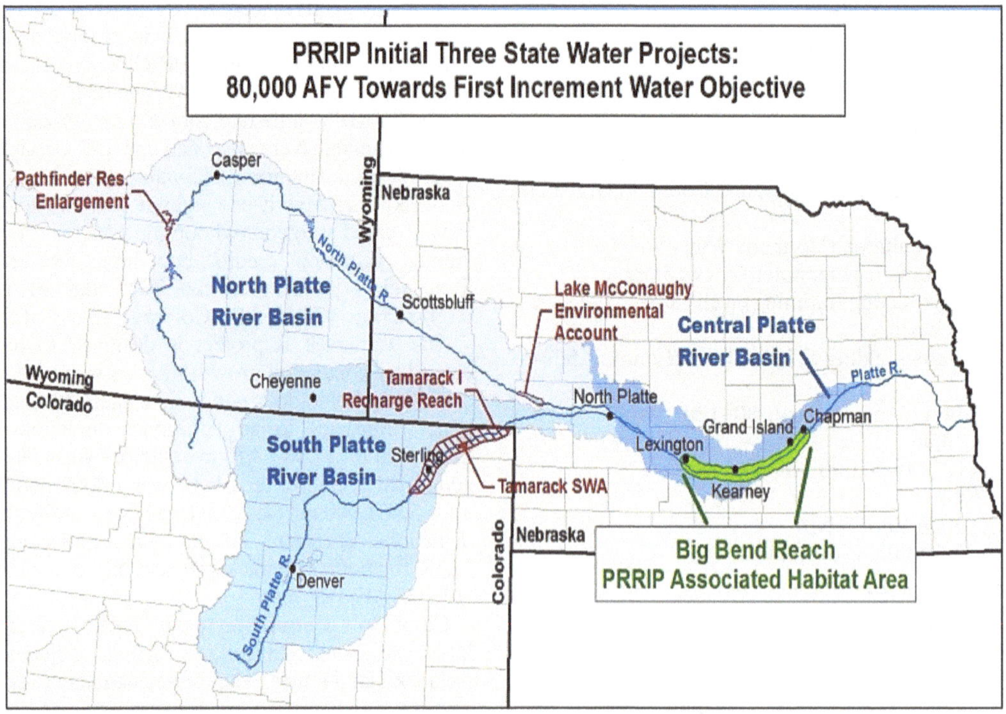

Figure D-5. Platte River Recovery Implementation Program initial water projects.
Source: Altenhofen (2012).

Figure D-6. Pictures from Tamarack Recharge and Wildlife project.
Source: Altenhofen (2012).

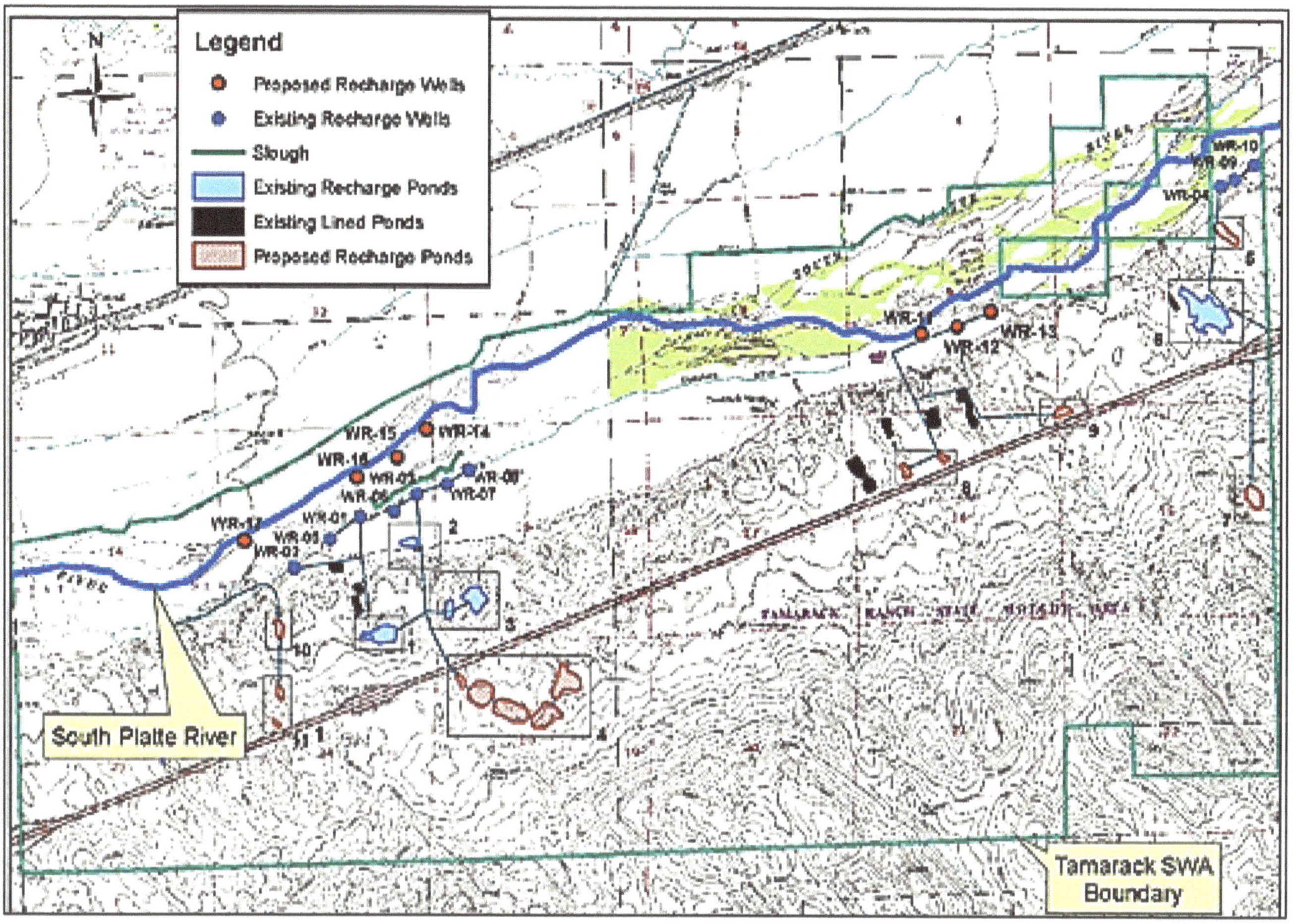

Figure D-7. Details of the Tamarack Recharge and Wildlife project.
Source: Altenhofen (2012).

Tamarack Ranch operations and maintenance costs of about \$68,000 per m^3 (\$55 per AF) of accretion credit. In addition to the Tamarack Ranch Project, Colorado water users also lease excess accretion credits for PRRIP from private existing augmentation and recharge plans in the lower South Platte River in Colorado.

Particular Challenges Encountered or Overcome. The plan is complicated because it involves multiple federal and state entities, multiple junior and senior water rights and court cases, compact issues, complicated infrastructure, and multiple endangered species.

There have been several hurdles to overcome since recharge began in 1996, including

- Water availability has been less than originally forecasted using historical data;
- Numerous water rights court cases;
- Groundwater modeling improvements to satisfy water rights objectors; and
- Continuous expansion of the number of pumping wells, recharge ponds, and monitoring wells.

Altenhofen (2012), Burns (1985), Halstead (2012), Shrier et al. (1999), Vail (2012), and Wallace (2012) provide details of this project.

Acknowledgments. Jon Altenhofen, P.E., Northern Water

Dennis McGrane, P.E., C.P.G., McGrane Water Engineering, LLC, 303-917-1247

CASE STUDY 5 DESCRIPTION

Project Name: Southern Avra Valley Storage and Recovery Project (SAVSARP)
Project Location (City/State): Tucson, Arizona
Operating Agency: Tucson Water
Project Contact: Dick L. Thompson, Lead Hydrologist
Phone: 520-837-2236, Email: Dick.Thompson@tucsonaz.gov
MAR Project Type: ☒ Surface Basin ☐ Subsurface ASR ☐ Vadose Zone Well
Basin Area (acres): 236
Well Depth (ft below ground surface): NA
Flow Rate (MGD): Variable
Recharge Rate: 30,000 hectare/yr (75,000 AF/yr), permitted maximum
Recovery Rate: 19,000 hectare/yr (48,000 AF/yr), equipment capability maximum
Year Completed: 2008
Currently Operational: ☒ Yes ☐ No
Cost (\$): \$73.3 million
Demonstration Phase? Yes ☐ / No ☒

Project Description. SAVSARP is the second of two projects that were constructed to recharge Tucson Water's 177,857,000 m^3

Figure D-8. The Southern Avra Valley Storage and Recovery Project recharge basins.
Source: City of Tucson, Arizona.

(144,191 AF) allocation of Colorado River water. The river water is delivered to Tucson by the Central Arizona Project (CAP). Test basins were used while the facility was under construction, and SAVSARP was brought to full operation in 2008 (Figure D-8). Including deliveries while under construction, SAVSARP has recharged 643,000,000 m^3 (521,000 AF) of water as of the end of 2016. After recharging, water is recovered by the project's 32 production wells and delivered to Tucson Water customers. The city of Tucson is the largest customer of CAP water delivery and operates a surface water basin recharge and recovery project in Arizona. In addition to its river allotment, Tucson Water also stores water for Arizona Water Bank Authority, Pascua Yaqui Tribe, Tohono O'odham Nation, the towns of Vail and Oro Valley, and the City of Phoenix.

Particular Challenges Encountered or Overcome: Nitrate Flushing. The land the SAVSARP facility is built on had a buildup of natural nitrates present in the desert Southwest and artificial deposition from the use of nitrogen fertilizers in cotton farming that had previously taken place on the property. Tucson Water has seen nitrate flushing occur in older recharge projects. Being well prepared, the utility instituted a private well sampling program for off-site wells and monthly sampling of facility monitoring wells close to the basins. The peak event documented was 12 mg/L of nitrate as nitrogen (N), with a duration of less than one month. Within two years of the onset of the flush, levels dropped below the original groundwater background readings. Amplitude and duration of the flushing appear to be directly related to the concentration of nitrate present in the vadose zone and the speed and volume of recharge water moving through the area. Six monitoring wells showed an average flushing event of 560 days, and 30 days were above the 10-part-per-million maximum contaminant level (MCL). By careful monitoring and mixing of recovered water during the flush, Tucson water customers were never affected.

Acknowledgments. Dick L. Thompson, Tucson Water
Margaret Snyder, Tucson Water

CASE STUDY 6 DESCRIPTION

Project Name: Montebello Forebay Spreading Grounds
Project Location (City/State): Pico Rivera, California
Operating Agency: Los Angeles County Department of Public Works (LADPW)
Project Contacts: Eric Batman (LADPW), Ted Johnson [Water Replenishment District of Southern California (WRD)]
Phone: 626-458-6308 (LADPW), 562-275-4240 (WRD)
Email: ebatman@dpw.lacounty.gov, tjohnson@wrd.org
MAR Project Type: ☒ Surface Basin ☒ In-Stream Check Dams ☐ Vadose Zone Well
Basin and In-Stream Area (acres): 834
Well Depths (ft below ground surface): NA
Basin Intake Capacity (cfs): 2,300
Years Completed: 1937–1955
Currently Operational: ☒ Yes ☐ No
Cost ($): N/A
Demonstration Phase? Yes ☐ / No ☒

Project Description. The Montebello Forebay Spreading Grounds (MFSG) include the San Gabriel Coastal and Rio Hondo Coastal surface spreading basins and the unlined portion of the San Gabriel River. The spreading grounds consist of 23 shallow spreading basins adjacent to the Rio Hondo and the San Gabriel River, seven inflatable rubber check dams in the San Gabriel River, effluent discharge locations and diversion structures, and an interconnection pipeline. The facilities are used to replenish groundwater in the over-drafted Central Groundwater Basin, which underlies a portion of the Los Angeles Coastal Plain. The project is owned and operated by the Los Angeles County Department of Public Works (LACDPW). The Water Replenishment District of Southern California (WRD) manages the groundwater basin and purchases water for replenishment to supplement natural recharge. The MFSG recharge facilities are shown in Figure D-9.

Figure D-9. Montebello Forebay Spreading Grounds recharge facilities.
Source: Courtesy of WRD.

Since the late 1930s, LACDPW has been using the MFSG to recharge stormwater runoff into the groundwater basin. Supplemental water sources outside of the basin (imported water) were secured in the 1950s to meet increasing groundwater demands and replenishment needs; recycled wastewater provided an additional water source for MFSG recharge in 1962. Because the long-term reliability of imported water supplies cannot be guaranteed, the WRD has recently developed the Groundwater Replenishment Improvement Program (GRIP) to replace imported water with more reliable recycled water for groundwater recharge at the MFSG.

As part of the GRIP, advance-treated recycled water is either directly injected into deep recharge wells or blended with tertiary water for spreading at the MFSG. The development of advance-treated recycled water eliminates the need for a diluent water source as required by California regulations for recharge of tertiary treated recycled water.

Particular Challenges Encountered or Overcome. Though advance-treated recycled water is generally considered a higher quality water source compared with tertiary treated recycled water, there are water quality considerations with respect to where and how the advance-treated recycled water is recharged. A study commissioned by WRD (Drewes et al. 2010) explored the effects of different blend ratios of tertiary and advance-treated recycled water on the quality of the underlying groundwater through controlled laboratory-scale experiments.

During soil aquifer treatment (SAT), dissolved organic carbon (DOC) in the tertiary treated recycled water provides a carbon source for microorganisms and results in increased biological attenuation and removal of certain wastewater constituents. Because the MFSG has been receiving recycled water since 1962, there is a considerable amount of total organic carbon (TOC) bound in the shallow subsurface soils. The study, which used soil columns to simulate SAT, indicated that using 100% advance-treated recycled water without blending with tertiary treated water had the potential to liberate the bound TOC into the shallow groundwater. An increase in nitrate concentration was also observed in the columns and attributed to the oxidizing of any remaining ammonia adsorbed to the soil matrix. Thus, spreading basins that have historically received secondary or tertiary treated recycled water for recharge may have water quality concerns should the source water be switched to advance-treated recycled water only. The study recommended a blending percentage of tertiary treated water with advance-treated recycled water to avoid the potential for TOC leaching.

To address this issue, WRD plans to blend the tertiary and advance-treated recycled water before recharge in the MFSG, thus optimizing SAT capabilities.

CASE STUDY 7 DESCRIPTION

Project Name: Denver Basin Aquifer Recharge Demonstration Project
Project Location: Denver, Colorado
Project Contact: Richard Clark, Sr. Engineering Advisor, East Cherry Creek Valley Water and Sanitation District
Phone: 303-693-3800, Email: RClark@ECCV.org
MAR Project Type: ☐ Surface Basin ☒ Subsurface ASR or Vadose Zone Well
Basin Area (acres): NA
Flow Rate (MGD): 1.4
Year Completed: 1996
Currently Operational: ☐ Yes ☒ No
Cost ($): $3,288,578
Demonstration Phase: ☐ Yes ☒ No

Project Description. The Willows Water District (WWD) project began in 1990 as one of 14 projects in the US Bureau of Reclamation's High Plains States Groundwater Demonstration Program. The project was performed to evaluate technical, institutional, legal, and economic issues associated with deep

Figure D-10. Site of Denver Basin Recharge Demonstration Project.

bedrock injection, storage, and recovery in the Denver Basin aquifer system. Between 1992 and 1996, 34 recharge and recovery cycles were performed at WWD Well A-6A, a 50 cm (20 in.) diameter, 474.2 m (1,556 ft) deep well with 20 screened intervals in the Arapahoe Aquifer. Well A-6A was installed as a production well in 1988 with a design capacity of 5,700 L/min (1,500 gpm); project recharge design rates of 4,500 L/min (1,200 gpm) were anticipated. The WWD ASR project was completed in fall 1996 after four years of testing, resulting in a net volume of 1,583,000 m^3 (1,283 AF) of injected water stored in the aquifer.

Well efficiency declined throughout the project and was attributed to the temperature-related dynamic viscosity of cold recharge water, particularly with respect to ambient groundwater pumped during pre-recharge aquifer testing. Data reviewed indicate that temperature-dependent viscosity was a minor component of decreased well efficiency and that microbial growth and adverse geochemical effects likely played a more significant role in well efficiency declines.

Particular Challenges Encountered or Overcome. In 1988, when Well A-6A was first installed (Figure D-10), specific capacity was 8.8 L/min/m (7.6 gpm/ft); this parameter had declined to 7.1 L/min/m (6.2 gpm/ft) during pre-recharge testing in 1991. After recharge start-up, specific capacity declined to less than 5.8 L/min/m (5 gpm/ft) and was below 3.5 L/min/m (3 gpm/ft) at project completion in 1996.

Thermal Effects. Well efficiency loss reported by the WWD was attributed to the higher viscosity of cold recharge water (1.6 °C to 14.8 °C) relative to native groundwater pumped for pre-recharge testing (21.8 °C). A correlation between water temperature and well efficiency decline was observed during recharge cycles using water of different temperatures (JCHA 1997). However, comparing theoretical well efficiency to observed efficiency indicates that temperature-dependent viscosity was not a substantial cause of well efficiency loss. As an example, after 10,000 min injecting at 2,002 L/min (529 gpm), comparing theoretical with observed heads indicates that of an observed 76 m (250 ft) of head buildup, approximately 9.1 m (30 ft) of head buildup can be attributed to temperature-dependent viscosity. After 1,000 min of pumping at 3,620 L/min (956 gpm), a similar comparison indicates that of an observed 79 m (260 ft) of drawdown only 9 m (30 ft) of drawdown can be attributed to temperature-dependent viscosity (JCHA 1994, Figures 6 and 11).

Microbial and Geochemical Effects. Iron-related bacteria (IRB) were observed in Well A-6A as purple or rust-colored deposits in a 1991 pre-recharge video log (Black and Veach 1993) and later in cultures from water samples obtained from the well (JCHA 1997). IRB derive their energy by oxidizing dissolved iron to insoluble iron hydroxides while producing a mucous-like biofilm and can thrive in cold, nutrient-poor water, particularly when mixing of aerobic water with anaerobic water containing dissolved iron occurs. Mass balance calculations indicate that over the program approximately 13.4 kg (29,500 lb) of oxygen and approximately 1.1 kg (2,500 lb) of chlorine were consumed by the aquifer (JCHA 1997). The dissolved iron concentration in the aquifer was up to 0.5 mg/L. This concentration indicates that significant oxidation of dissolved iron was possible in the aquifer and that chlorine likely oxidized bacterial biomass. Dissolved oxygen introduced into the aquifer appears to have stimulated previously low levels of IRB growth during the first recharge cycle when a substantial decline in specific capacity was observed. Repeated cycles of recharge are expected to have transported bacteria further from the well bore as water flushed back and forth over a growing zone of infection with each recharge cycle. During well rehabilitation, continuous gravel pack behind blank casing sections among the well's 20 screened intervals would not be treated effectively and could provide a reservoir for bacterial reinfection.

Responding to continuing well efficiency declines and observed IRB in samples from the well, rehabilitation was attempted by injecting liquefied CO_2 into the screened zones in the fall of 1993. Specific capacity increased from 3.7 L/min/m (3.2 gpm/ft) to 4.4 L/min/m (3.8 gpm/ft) yet continued to decline afterward. Liquid CO_2 recharge likely exacerbated conditions by causing calcium carbonate precipitation near the well bore by the following process. The injected CO_2 would produce carbonic acid. At low pH, hydrogen ions would displace adsorbed Ca^{2+} ions from the aquifer matrix, resulting in a pH rebound. Increased pH would then decrease calcite solubility while increasing groundwater Ca^{2+} concentrations. This process would leave water charged with Ca^{2+} and HCO_3^- ions, resulting in calcite precipitation. To compound matters, cultures of a common IRB, *Gallionella ferruginea*, have been observed to grow autotrophically (synthesizing its own food from inorganic substances) using CO_2 as the sole carbon source (Hallbeck and Pederson 1986).

The 50% decline in well efficiency observed at Well A-6A highlights the importance of detailed pre-recharge testing, including microbiological and geochemical evaluation before well selection and during ASR operation and maintenance.

More information on this case study can be found in Black and Veach (1993), Hallbeck and Pederson (1986), and John C. Halapaska and Associates et al. (1994, 1997).

Acknowledgments. Bruce Lytle and Khanh Le generously provided their insights and access to project reports and unpublished data during the research for this evaluation, which was prepared by Andy Horn.

CASE STUDY 8 DESCRIPTION

Project Name: OCWD Talbert Gap Seawater Intrusion Barrier
Project Location (City/State): Fountain Valley, California
Operating Agency: Orange County Water District
Project Contacts: John Bonsangue and Tim Sovich
Phone: 714-378-3353 (JB), 714-378-3226 (TS)
Email: jbonsangue@ocwd.com, tsovich@ocwd.com
MAR Project Type: ☐ Surface Basin ☒ Subsurface
☐ Vadose Zone Well
Basin Area (acres): NA
Well Depths (ft below ground surface): 100–700
Flow Rate (MGD): 20–36
Year Completed: 1975
Currently Operational: ☒ Yes ☐ No
Cost ($): NA
Demonstration Phase? Yes ☐ / No ☒

Project Description. The Orange County Groundwater Basin is largely protected from seawater intrusion by the Newport–Inglewood Fault zone, which cuts off basin aquifers from the ocean. However, near the coast, seawater intrusion can occur in gaps in the Newport–Inglewood Fault zone. The Talbert Gap in Fountain Valley is an erosional feature cut by the Santa Ana River, which deposited permeable sediments in the gap, thereby creating pathways for seawater to bypass the fault zone and enter basin aquifers. Seawater intrusion in the gap was first noticed in the 1920s. To minimize further seawater intrusion, the Talbert Barrier was constructed and went online in the mid-1970s. The Talbert Barrier consists of 36 recharge well sites that recharge multiple aquifer zones to prevent seawater intrusion and to replenish the groundwater basin (Figure D-11).

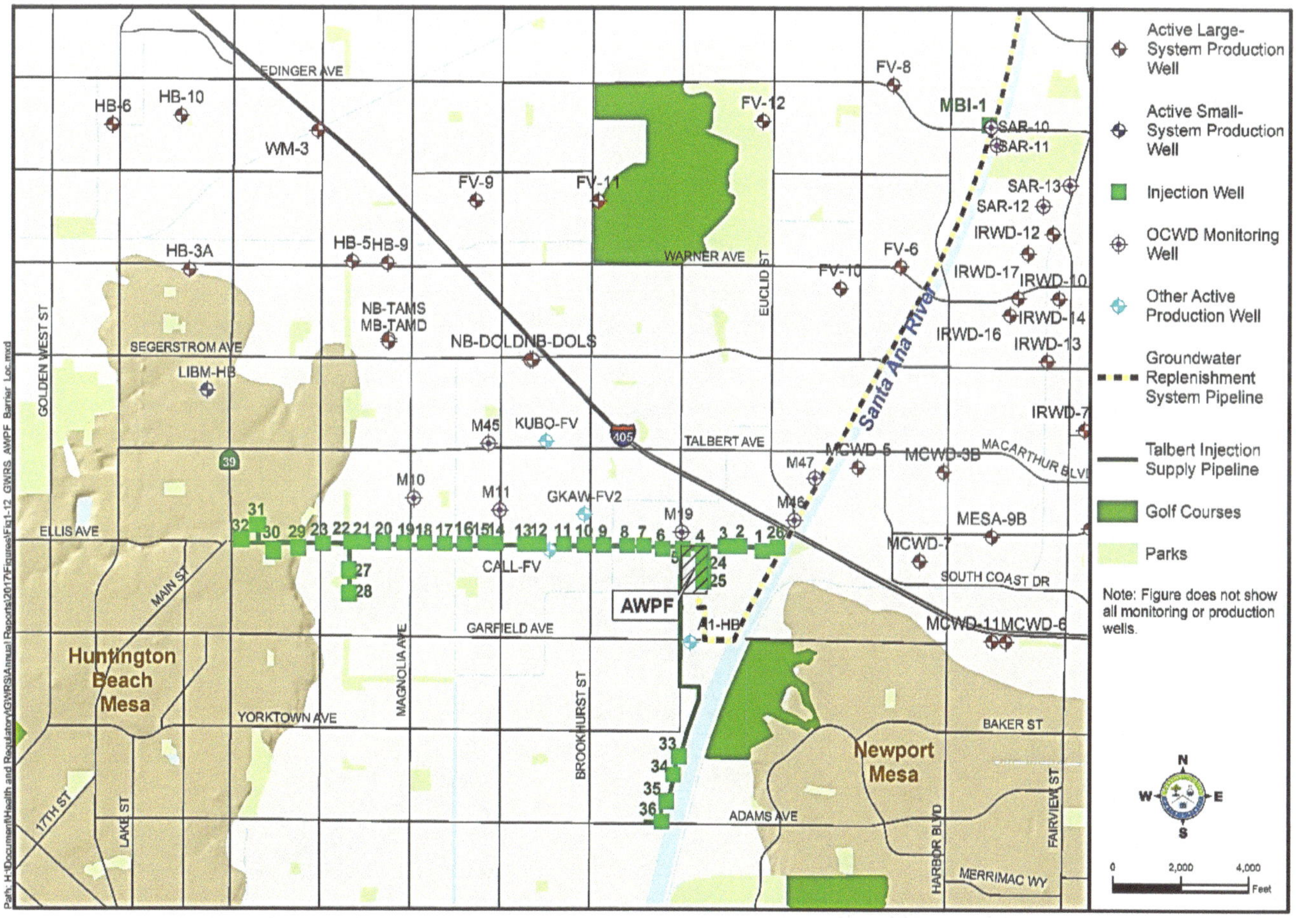

Figure D-11. Overview of the Talbert Gap seawater barrier project.
Source: Orange County Water District.

From the mid-1970s to 2008, the source water for the barrier was a mixture of recycled water from Water Factory 21, imported water from the Metropolitan Water District of Southern California, and amber-colored deep well water. Because of the organic content in the colored deep well water, clogging was a constant issue that had to be addressed with regular well redevelopment. Annual barrier capacity during this period ranged from 3 to 16 MGD. In 2008, with the completion of OCWD's Groundwater Replenishment System (GWRS), which included the addition of several new recharge wells, the barrier now receives 100% recycled water treated with microfiltration, reverse osmosis, and advanced oxidation. Because of the barrier expansion, superior water quality, and improved barrier operations, such as more frequent backflushing and rehabilitation, barrier capacity now typically ranges from 20 to 36 MGD. Current barrier capacity largely depends on overall basin storage conditions and seasonally varying coastal water levels.

Particular Challenges Encountered or Overcome. Even with purified recycled water as a source water, clogging because of particulates derived mostly from degradation of the conveyance pipelines remains a challenge. A regular redevelopment program is implemented to backflush or redevelop the wells at regular intervals to maintain capacities and prevent any permanent clogging.

Also, California regulations require a 3- to 6-month travel time buffer area to be established around the barrier in which no potable groundwater production wells can be located.

CASE STUDY 9 DESCRIPTION

Project Name: City of Phoenix Aquifer Storage and Recovery (ASR) Well 299
Project Location (City/State): Phoenix, Arizona
Operating Agency: City of Phoenix
Project Contact: Aimee Conroy
Phone: 602-495-5654, Email: aimee.conroy@phoenix.gov

MAR Project Type: ☐ Surface Basin ☒ Subsurface ASR
☐ Vadose Zone Well
Basin Area (acres): NA
Well Depth (ft below ground surface): 1,420
Flow Rate: 9.4 to 9.8 × 10^6 L/d (2.5 to 2.6 MGD) recharge; 1.3 × 10^7 L/d (3.4 MGD) recovery
Year Completed: 2010
Currently Operational: ☒ Yes ☐ No
Cost ($): $3.8 million
Demonstration Phase? Yes ☐ / No ☒

Project Description. Well 299 was not only the first ASR well to be installed by the City of Phoenix, but also the first ASR well in the United States to incorporate the reverse siphon recharge method and deliver recovered supplies directly into the distribution system. The reverse siphon method uses the well pump and aboveground valving to initiate recharge without entraining air in the filter pack and formation—without the use of a subsurface flow control valve (Figure D-12). Entrainment of air into the aquifer causes increased clogging by reducing pore spaces in the filter pack and formation (semi-consolidated and unconsolidated sediments). The pump is started to purge the column pipe and borehole of loose particulates and expel more than 170 m (550 ft) of air from the column pipe to the atmosphere. When the water clears, the recharge valve allows water to flow from the distribution system to the wellhead. Because the pump is still running, the distribution system water is forced to the pump-to-waste line, mixing with the already flowing pumped groundwater. When the recharge valve is fully open, the well pump shuts down. A large portion of the recharge flow immediately begins recharging into the well, while the remaining portion continues to flow to the pump-to-waste tank. The pump-to-waste valve then closes and terminates flow to the pump-to-waste tank. This method of recharge has been successful during the first three years of operation and has undergone more than 275 operation cycles.

Figure D-12. City of Phoenix ASR Well 299.

Accomplishments within three years of operations include the following:

- Recharge rates of 6,541–7,086 L/m (1,728–1,872 gpm) have consistently performed at 72% to 78% of the recovery rate (9085 L/m; 2,400 gpm);
- The recharge rate is consistently 9.4–9.8 million L/day (2.5 to 2.6 MGD);
- The recharge rate has dropped less than 2% over three years of operations;
- The recovery rate 9,100 L/m (2,400 gpm) has remained consistent;
- Reduction of downhole components allows operators to observe and maintain all mechanical equipment associated with well activity aboveground;
- A dechlorination system was developed for recharge operations to reduce the potential of disinfection by-products (DBPs) in the aquifer; and
- Well 299 provides the ability to backwash/unclog the well with the existing well pump assembly, saving the city $110,000 to $150,000 annually in well rehabilitation costs.

Particular Challenges Encountered or Overcome. The complex well design was carefully developed to address multiple technical, logistical, and regulatory requirements, including the need to minimize clogging during recharge cycles in an effective and cost-efficient manner. The Well 299 challenges included the following:

- Creating an operator-friendly system that is efficient and cost-effective,
- Engineering a well system that allows the well to pump directly into the distribution system and meet regulatory requirements, and
- Maintaining consistent recharge and recovery rates over time.

Performance accomplishments at Well 299 include the following:

- Reduces well rehabilitation costs by $110,000 to $150,000 per year (three years of operational savings total $330,000 to $450,000);
- Reduces well rehabilitation work duration from 30–40 days to 3–4 days;
- Operator-friendly programming system has enhanced the operations and management of the ASR well system (reduced labor force from 2–3 operators to 1 operator);
- Uses the permanent pump assembly, which can unclog the well after subsequent recharge cycles; and
- Improves the lifecycle of the ASR well system [from $422,000/m^3 to $249,000/m^3 ($342/AF to $202/AF)].

More information on this case study can be found in Gin et al. (2013, 2011), Gin and Wrzosek (2013), and Gin and Nutter (2011, 2012).

Acknowledgments. Marvin Glotfelty, R.G., Clear Creek Associates
Dave Wrzosek, R.G., Clear Creek Associates
Nathan E. Nutter, P.E., Carollo Engineers
Tom Morris, ASR Systems
Jeff Wold, Weber Water Resources LLCt

CASE STUDY 10 DESCRIPTION

Project Name: Rio Rancho Direct Injection Demonstration Project
Project Location (City/State): Rio Rancho, New Mexico
Operating Agency: City of Rio Rancho
Project Contact: Jim Chaisson, P.E., Utilities Division Manager
Phone: 505-896-8715, Email: jchiasson@rrnm.gov
MAR Project Type: ☐ Surface Basin ☒ Subsurface ASR ☐ Vadose Zone Well
Basin Area (acres): NA
Well Depth (ft below ground surface): 1,700
Flow Rate (MGD): 1
Year Completed: 2013
Currently Operational: ☒ Yes ☐ No
Cost ($): $4.0 million
Demonstration Phase? Yes ☐ / No ☒

Project Description. Innovative water supply alternatives, such as water reuse coupled with aquifer replenishment, are needed to support the city of Rio Rancho's projected growth within this water-limited region. The city has demonstrated that direct recharge methods can be used to safely replenish the underlying aquifer with a purified, reclaimed water source. This replenishment includes the design, permitting, and development of the first direct recharge facility in the state of New Mexico, consisting of one deep recharge well [0.43 m (16 in.) casing to 520 m (1,700 ft)], surrounded by an extensive groundwater monitoring network. The direct recharge facility is capable of replenishing the groundwater system at a rate of 378,540 L/day (1 MGD).

Figure D-13. Rio Rancho Direct Injection Demonstration Project facilities.

The city has obtained discharge permits through the New Mexico Environment Department (NMED) and Underground Storage and Recovery permits from the New Mexico Office of the State Engineer (OSE) for operation of the aquifer recharge facility. Permitting of recharge operations with a purified water source required an extensive evaluation of treatment options to determine best available technology that protects public health and the water resource. Screening of treatment technologies has ranged from physical separation through reverse osmosis (RO) membranes, advanced oxidation processes using ozone and hydrogen peroxide, adsorption processes [granular activated carbon (GAC) and biologically activated carbon (BAC)], and final disinfection. The city constructed an advanced water treatment facility that uses an ozone-based advanced oxidation process (AOP), followed by a BAC–GAC step to reduce remaining wastewater-derived organics in the membrane bioreactor (MBR) effluent before recharge within the regional aquifer (Figure D-13). This process has been used extensively for drinking water treatment and has shown great potential for reclaimed waters when coupled with MBR systems.

Particular Challenges Encountered or Overcome. Challenges include establishment of reasonable expectations for project feasibility, benefits, time to implement, costs, and a host of technical issues for aquifer replenishment with a reclaimed water source. These ongoing challenges are overcome through the efforts of dedicated city staff and consultants responsible for implementing the water reuse and aquifer replenishment strategic plan.

More information on this case study can be found in Hahn et al. (2003), Knorr and Cliett (1985), and Sheng (2005).

Acknowledgments. Robert Marley, former Project Manager/Senior Hydrogeologist, Daniel B. Stephens & Associates, Inc.

Gwinn Hall, P.E., Project Engineer, Daniel B. Stephens & Associates, Inc.

CASE STUDY 11 DESCRIPTION

Project Name: City of El Paso Clean Water Recharge
Project Location (City/State): El Paso, Texas
Operating Agency: El Paso Water
Project Contact: Scott Reinert, Water Resources Manager
Phone: 915-594-5628, Email: sreinert@epw.org
MAR Project Type: ☐ Surface Basin ☒ Subsurface ASR
☐ Vadose Zone Well
Basin Area (acres): NA
Well Depth (ft below ground surface): 10 wells
Flow Rate (MGD): total up to 10 (recharge); 0 (recovery)
Year Completed: 1984
Currently Operational: ☒ Yes ☐ No
Cost ($): $26.7 million for both Fred Hervey Reclamation Plant and 10 recharge wells
Demonstration Phase? Yes ☐ / No ☒

Project Description. The El Paso reclaimed wastewater MAR system was constructed and put into operation in 1985. Its source water is reclaimed wastewater treated at the El Paso Water Utilities (EPWU) Fred Hervey Water Reclamation Plant. Its storage space is the Hueco Bolson aquifer in northeast El Paso, where the geological formation is thick enough for effective injection and sufficient land access for effective recovery of the injected water is available. EPWU's MAR facilities include 10 recharge wells constructed in 1984 and one recharge basin built in 2001 with a delivery pipeline from the Fred Hervey Water Reclamation Plant (Figure D-14); its recovery facilities are neighboring production wells, primarily production wells that are downgradient of recharge wells, connected with a potable water distribution network.

This MAR system serves several purposes: first, storing the excess reclaimed wastewater and recovering it for future supply or to meet peak demand; second, partially restoring the depleted aquifer by raising its water level; and third, slowing down or preventing intrusion of brackish water. Since its start-up in 1985, the Fred Hervey Plant has produced approximately 121 million m^3 (32.1 billion gal.) of reclaimed wastewater, of which two thirds [74.7 million m^3 (19.7 billion gal.)] have been injected into the Hueco Bolson aquifer and approximately one third has been used for other purposes. The annual recharge peaked at 7.0 million m^3 (1.9 billion gal.) in 1990; however, the annual recharge rate has been reduced thereafter because of increases in demands for other uses of reclaimed water, primarily for cooling purposes at El Paso Electric Company. Recharge currently accounts for only approximately 35% to 50% of total produced

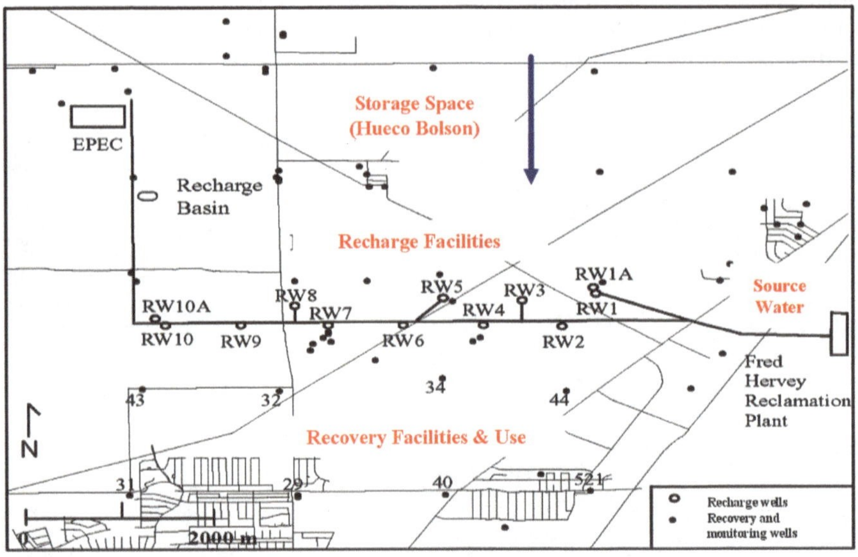

Figure D-14. City of El Paso Clean Water Recharge Project facility layout.
Source: Sheng (2005).

reclaimed wastewater. The Fred Hervey Plant was expanded from 37,900 m^3/d (10 MGD) to 45,400 m^3/d (12 MGD) in 2012.

Particular Challenges Encountered or Overcome. Regulatory requirements by the Texas Department of Health include a residence time of two years of the stored water to be recovered by the neighboring production wells. To meet this requirement, the recharge wells were designed to be located at least 600 m (2,000 ft) apart from the production wells.

Four recharge wells (RW1, 2, 3, and 10) have failed. The reasons for the failure include electrochemical corrosion caused by the foreign electric potential from existing cathodically protected gas pipelines, where the galvanized recharge line and pump columns serve as one anode and the well screen as another. The chemical content of the injected water (containing a lower pH with dissolved oxygen, higher chloride and sulfate concentrations than native groundwater, and free chlorine) and inadequate well development have also been the causes of failure for those wells. The following measures have been taken to prevent such failure. First, cathodic protection was installed by removing the recharge column and replacing the existing pump column with a black steel pipe to avoid electrochemical corrosion at some well locations. After the removal of the recharge column, the reclaimed water is injected through the pump column rather than through the recharge line. Second, recharge wells were developed regularly to maintain well screens in good condition and retain a good recharge capacity. Third, PVC or stainless-steel well screens were installed in new wells to avoid corrosion by chemical reaction or other factors instead of carbon steel screens. With the modification to the existing well system and improved design for new wells, the recharge system maintains good operational conditions.

Since 2001, EPWU gradually switched its recharge from recharge wells to infiltration basins (Hahn et al. 2003). EPWU currently maintains and operates six infiltration basins and two recharge wells.

More information on this case study can be found in Eaton et al. (2009) and GSI Water Solutions (1999 to 2016).

Acknowledgments. Scott Reinert, El Paso Water Utilities
Make Fahy, El Paso Water Utilities

CASE STUDY 12 DESCRIPTION

Project Name: City of Beaverton Aquifer Storage and Recovery (ASR) Program
Project Location (City/State): Beaverton, Oregon
Operating Agency: City of Beaverton
Project Contacts: David Winship, P.E., and Brion Barnett, P.E.
Phone: 503-526-2434 (DW), 503-526-2592 (BB)
Email: dwinship@ci.beaverton.or.us,
bbarnett@ci.beaverton.or.us

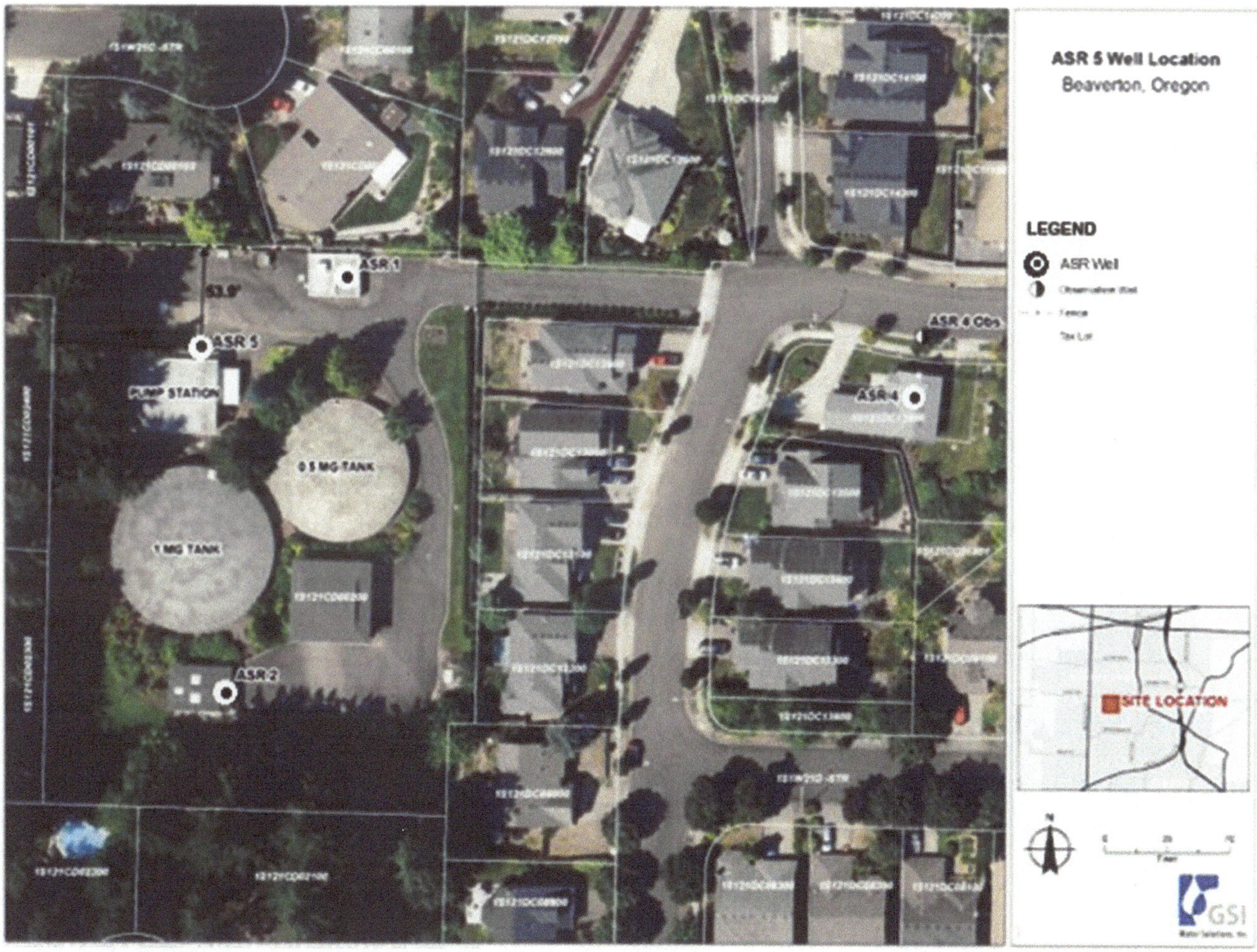

Figure D-15. Beaverton Aquifer Storage and Recovery well field.
Source: GSI Water Solutions (2016).

MAR Project Type: ☐ Surface Basin ☒ Subsurface ASR
☐ Vadose Zone Well
Basin Area (acres): NA
Well Depth (ft below ground surface): 500
Flow Rate: 5.3 million L/d (1.4 MGD) or more (recharge), 3 million L/d (8 MGD) (recovery) (peak using both wells)
Year Completed: 2001 (ASR 2), 2005 (ASR 4)
Currently Operational: ☒ Yes ☐ No
Cost ($): $4.7 million (2017 dollars)
Operational Phase? Yes ☒ / No ☐

Project Description. Despite the Pacific Northwest's reputation for being "wet," many cities west and east of the Cascade Range in the United States find it increasingly difficult to meet peak water supply demand during the dry summer months. Aquifers in the east side of Oregon and Washington are the primary water supply sources for a vast agriculture industry, and they have experienced significant declines, prompting regulatory restrictions. For these reasons, municipalities west and east of the Cascade Range, as well as agricultural interests, have opted to implement ASR projects as a unique water management technique to help meet peak summer water demands.

Unique to the Pacific Northwest are the Miocene Age continental flood–basalt flows of the Columbia River Basalt Group (CRBG), which consist of a thick, areally extensive series of extraordinarily huge lava flows. The CRBG plays host to an extensive regional aquifer system in eastern Washington, eastern Oregon, and western Oregon. The City of Beaverton ASR project uses the CRBG aquifers to host its ASR project. Specifically, the majority of water yield from the CRBG is from interflow zones that represent the top and bottom of each sheet flow; these interflow zones are highly productive and have limited vertical connection.

Since 1999, the city of Beaverton, Oregon, population 95,500, has installed three ASR wells hosted in the CRBG aquifer: ASR-1, ASR-2, and ASR-4; these three ASR wells are located at the Sorrento Water Works facility in Beaverton, and wells are no more than 120 m (400 ft) apart (Figure D-15). Despite their close proximity, interference is manageable because of the highly productive aquifer in the CRBG that is used to host the city's ASR wells. ASR-1 was recently decommissioned, because the well was drilled in 1945 and the pump system was old and failing. ASR-1 was originally retrofitted as the city's first ASR well at Sorrento in 1998. ASR-1 will be replaced at the same general location with a new ASR well (ASR-5). ASR-2 and ASR-4 use the upper 150 m (500 ft) of the CRBG section at this location, whereas ASR-5 uses the lower section of the CRBG from 150–300 m (500–1,000 ft) belowground surface (Figures D-15 and D-16). In short, this will be a stacked ASR system, where vertical hydraulic connection through secondary fractures is magnitudes less compared with the horizontal hydraulic permeability of the interflow zones. Currently, the city has the capacity to store approximately 1.7 million m^3 (up to 450 million gal.) of treated drinking water annually using its ASR wells at Sorrento. The three wells can provide up to about 30,500 m^3/day (8 million gal./day) at peak capacity, which is equivalent to at least 35% of the city's summer peak day demand. A cost–benefit analysis of the Beaverton ASR system

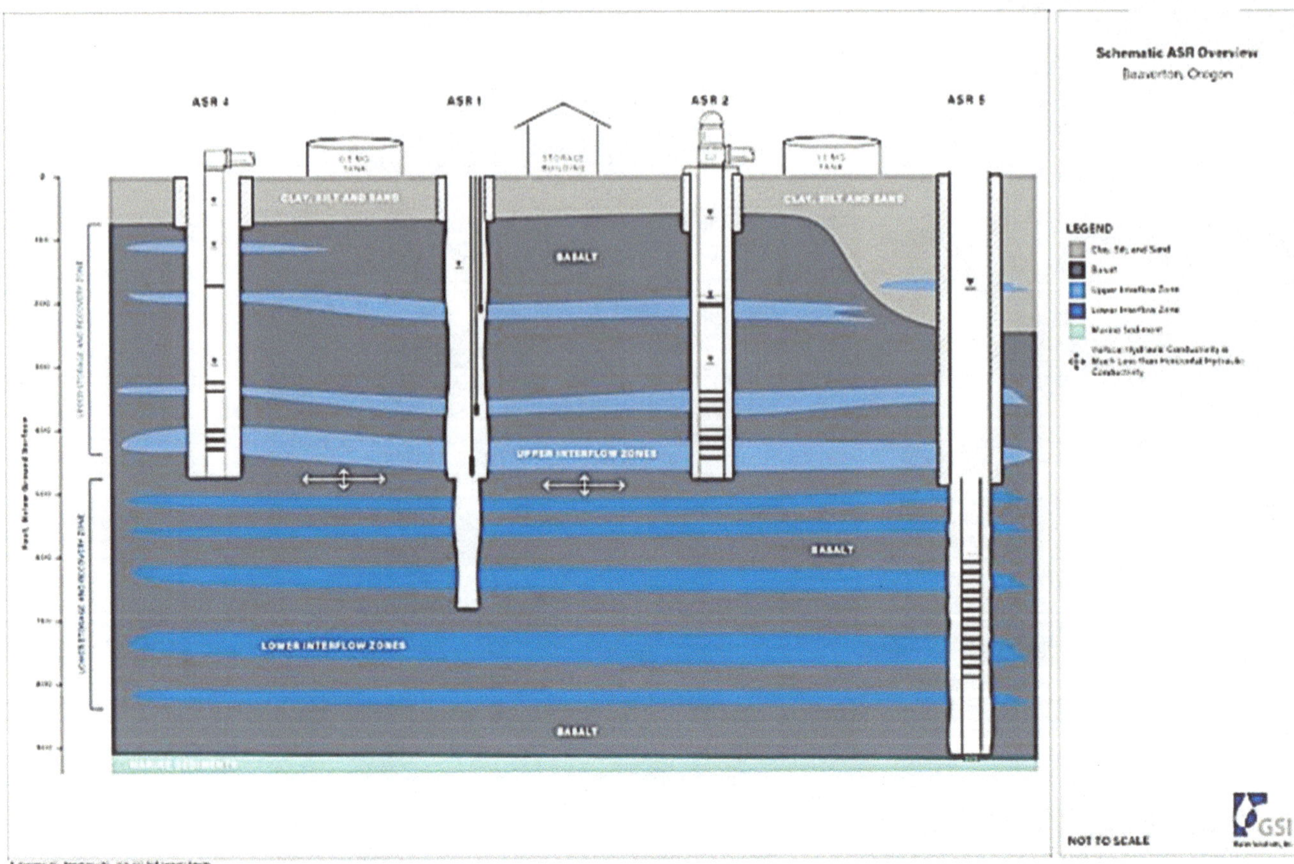

Figure D-16. Beaverton Aquifer Storage and Recovery well field cross-section
Source: GSI Water Solutions (2016).

relative to a conventional water supply upgrade (e.g., increasing transmission size capacity, reservoir upgrades, or treatment capacity upgrades, including O&M) completed in 2009 showed that ASR for the city has a benefit/cost ratio of 1.26 over a conventional source. Favorable hydrogeologic response and significant economic savings have made the city's ASR system immensely successful.

Particular Challenges Encountered or Overcome. Challenges encountered with the Beaverton ASR system include drilling, managing water during aquifer testing, and constructing complex ASR wells and pump stations in a highly urbanized location with limited access. A robust outreach program, including close construction monitoring, is needed to ensure that the neighborhood is not unduly affected during drilling and construction. Design of the pump stations should try to blend into the urban environment; engaging the neighborhood early in the design process has proven to be valuable.

Key lessons learned for the Beaverton ASR project include the fact that storage in basalt is highly successful; understanding well and aquifer hydraulics in CRBG aquifers, however, is challenging because of their unique geologic characteristics (e.g., tabular interflows and compartmentalization); aquifer clogging by air entrainment is a concern but can be managed with the installation of a downhole control valve; recharge well design is important; radon dissolves quickly into stored water; surface water has proven to be geochemically compatible with native CRBG groundwater; continuous monitoring of recharge linked to a downhole control valve has proven to be successful; and ASR has proven to be a cost-effective peak water management tool for the city of Beaverton.

More information on this case study can be found in Eaton et al. 2009 and GSI Water Solutions, Inc. (1999 through 2016).

Acknowledgments. Larry G. Eaton, RG, LHG, GSI Water Solutions, Inc.

Jason Melady, RG, CWRE, GSI Water Solutions, Inc.

REFERENCES

Abu-Zreig, M., M. Attom, and N. Hamasha. 2000. "Rainfall harvesting using sand ditches in Jordan." *Agr. Water Manage.* 46 (2): 183–192.

Ahmed, M. M., and M. A. Ayoub. 2014. "A comprehensive study on the current pressure drop calculation in multiphase vertical wells; Current trends and future prospective." *J. Appl. Sci.* 14 (23): 3172–3181.

Altenhofen, J. 2012. "Colorado division of water resources irrigationists symposium, 2012." Colorado's Tamarack Plan of Managed Groundwater Recharge: Colorado's Water Contribution to 3-States ESA PRRIP. Accessed March 15, 2012. http://dwrweblink.state.co.us/dwrweblink/ElectronicFile.aspx?docid=2810657&&dbid=0.

Amy, G., and J. Drewes. 2007. "Soil aquifer treatment (SAT) as a natural and sustainable wastewater reclamation/reuse technology: Fate of wastewater effluent organic matter (EfOM) and trace organic compounds." *Environ. Monit. Assess.* 129 (1–3): 9–26.

Anderson, M. P., W. W. Woessner, and R. J. Hunt. 2015. *Applied groundwater modeling,* 2nd ed. Amsterdam, Netherlands: Elsevier.

ANSS-WGD (Advanced National Seismic System Technical Integration Committee-Working Group D). 2007. "Instrumentation guidelines for the advanced national seismic system." Prepared for the US Geological Survey and ANSS National Implementation Committee. Accessed February 15, 2013. http://earthquake.usgs.gov/monitoring/anss/docs/ANSS_WGD_InstrGuideline_June2007.pdf.

Aquifer Storage Recovery Technical Advisory Group. 2002. *A review of aquifer storage recovery techniques.* Madison, WI: Wisconsin Dept. of Natural Resources.

ASCE. 1996. *Hydrology handbook: ASCE manuals and reports on engineering practice no. 28,* 2nd ed., 281–293. Reston, VA: ASCE.

ASCE. 2001. *Standard guidelines for artificial recharge of ground water.* ASCE/EWRI 34. Reston, VA: ASCE.

ASCE. 2009. *Comprehensive transboundary water quality management agreement, with guidelines for development of a management plan, standards, and criteria.* ASCE/EWRI 33. Reston, VA: ASCE.

ASCE. 2010. *Guidelines for the physical security of water utilities.* ANSI/ASCE/EWRI 56. Reston, VA: ASCE.

ASCE. 2012. *Guideline for development of effective water sharing agreements.* ASCE/EWRI 60. Reston, VA: ASCE.

ASCE. 2016. *ASCE rules for standards committees.* Reston, VA: ASCE.

ASCE. 2018. *Regulated riparian model water code.* ASCE/EWRI 40. Reston, VA: ASCE.

ASTM. n.d. *Guide for application of a ground-water flow model to a site-specific problem.* ASTM D5447. West Conshohocken, PA: ASTM.

ASTM. n.d. *Guide for comparing ground-water flow model simulations to site-specific information.* ASTM D5490. West Conshohocken, PA: ASTM.

ASTM. n.d. *Guide for comparison of field methods for determining hydraulic conductivity in the vadose zone.* ASTM D5126. West Conshohocken, PA: ASTM.

ASTM. n.d. *Guide for conducting sensitivity analysis for a ground-water flow model application.* ASTM D5611. West Conshohocken, PA: ASTM.

ASTM. n.d. *Guide for defining boundary conditions in ground-water flow modeling.* ASTM D5609. West Conshohocken, PA: ASTM.

ASTM. n.d. *Guide for defining initial conditions in ground-water flow modeling.* ASTM D5610. West Conshohocken, PA: ASTM.

ASTM. n.d. *Guide for development of ground-water monitoring wells in granular aquifer.* ASTM D5521. West Conshohocken, PA: ASTM.

ASTM. n.d. *Guide for displaying results of chemical analyses of groundwater for major ions and trace elements-diagrams for single analyses.* ASTM D5738. West Conshohocken, PA: ASTM.

ASTM. n.d. *Guide for displaying results of chemical analyses of groundwater for major ions and trace elements—trilinear and other multiple coordinate diagrams.* ASTM D5754. West Conshohocken, PA: ASTM.

ASTM. n.d. *Guide for displaying the results of chemical analyses of groundwater for major ions and trace elements—diagrams based on data analytical calculations.* ASTM D5877. West Conshohocken, PA: ASTM.

ASTM. n.d. *Guide for documenting a ground-water flow model application.* ASTM D5718. West Conshohocken, PA: ASTM.

ASTM. n.d. *Guide for planning and conducting borehole geophysical investigations.* ASTM D5753. West Conshohocken, PA: ASTM.

ASTM. n.d. *Guide for planning and preparing for a ground water sampling event.* ASTM D5903. West Conshohocken, PA: ASTM.

ASTM. n.d. *Guide for pore-liquid sampling form the vadose zone.* ASTM D4696. West Conshohocken, PA: ASTM.

ASTM. n.d. *Guide for soil sampling from the vadose zone.* ASTM D4700. West Conshohocken, PA: ASTM.

ASTM. n.d. *Guide for use of cable tool drilling and sampling methods for environmental exploration and installation of subsurface water quality monitoring devices.* ASTM D5875. West Conshohocken, PA: ASTM.

ASTM. n.d. *Guide for use of direct air rotary drilling for environmental exploration and installation of subsurface water quality monitoring devices.* ASTM D5782. West Conshohocken, PA: ASTM.

ASTM. n.d. *Guide for use of dual well-reverse circulation-drilling for environmental exploration and installation of subsurface water quality monitoring devices*. ASTM D5781. West Conshohocken, PA: ASTM.

ASTM. n.d. *Guide for using seismic refraction method for subsurface investigation*. ASTM D5777. West Conshohocken, PA: ASTM.

ASTM. n.d. *Guide to site characterization for engineering design and construction purposes*. ASTM D420. West Conshohocken, PA: ASTM.

ASTM. n.d. *Practice for design and installation of ground water monitoring wells in aquifers*. ASTM D5092. West Conshohocken, PA: ASTM.

ASTM. n.d. *Standard guide for calibrating a groundwater flow model application*. ASTM D5981. West Conshohocken, PA: ASTM.

ASTM. n.d. *Standard guide for displaying the results of chemical analyses of groundwater for major ions and trace elements — use of maps*. ASTM D5410. West Conshohocken, PA: ASTM.

ASTM. n.d. *Standard guide for environmental site characterization in cold regions*. ASTM D5730. West Conshohocken, PA: ASTM.

ASTM. n.d. *Standard guide for selection of aquifer test method in determining hydraulic properties by well techniques*. ASTM D4043. West Conshohocken, PA: ASTM.

ASTM. n.d. *Standard guide for selection of data elements for groundwater investigations*. ASTM D5474. West Conshohocken, PA: ASTM.

ASTM. n.d. *Standard guide for set of data elements to describe a groundwater site; part one—Additional identification descriptors*. ASTM D5408. West Conshohocken, PA: ASTM.

ASTM. n.d. *Standard guide for set of data elements to describe a groundwater site; part two—Physical descriptors*. ASTM D5409. West Conshohocken, PA: ASTM.

ASTM. n.d. *Standard practice for determining the specification conformance of geosynthetics*. ASTM D4759. West Conshohocken, PA: ASTM.

ASTM. n.d. *Standard practice for (field procedure) for constant drawdown tests in flowing wells for determining hydraulic properties of aquifer systems*. ASTM D5786. West Conshohocken, PA: ASTM.

ASTM. n.d. *Standard practice for minimum set of data elements to identify a groundwater site*. ASTM D5254. West Conshohocken, PA: ASTM.

ASTM. n.d. *Standard terminology relating to soil, rock, and contained fluids*. ASTM D653. West Conshohocken, PA: ASTM.

ASTM. n.d. *Standard test method for (analytical procedure) for determining transmissivity and storage coefficient of nonleaky confined aquifers by the modified Theis nonequilibrium method*. ASTM D4105. West Conshohocken, PA: ASTM.

ASTM. n.d. *Standard test method for (analytical procedure) for determining transmissivity and storage coefficient of nonleaky confined aquifers by the Theis nonequilibrium method*. ASTM D4106. West Conshohocken, PA: ASTM.

ASTM. n.d. *Standard test method for centrifuge moisture equivalent of soils*. ASTM D425. West Conshohocken, PA: ASTM.

ASTM. n.d. *Standard test method for determining transmissivity of nonleaky confined aquifers by the Theis recovery method*. ASTM D5269. West Conshohocken, PA: ASTM.

ASTM. n.d. *Standard test method for infiltration rate of soils in field using double-ring infiltrometer*. ASTM D3385. West Conshohocken, PA: ASTM.

ASTM. n.d. *Standard test method for one-dimensional consolidation properties of soils*. ASTM D2435. West Conshohocken, PA: ASTM.

ASTM. n.d. *Standard test method for (field procedure) for withdrawal and injection well testing for determining hydraulic properties of aquifer systems*. ASTM D4050. West Conshohocken, PA: ASTM.

ASTM. n.d. *Standard test methods for measurement of hydraulic conductivity of saturated porous materials using a flexible wall permeameter*. ASTM D5084. West Conshohocken, PA: ASTM.

ASTM. n.d. *Test method (analytical procedure) for analyzing the effects of partial penetration of control well and determining the horizontal and vertical hydraulic conductivity in a nonleaky confined aquifer*. ASTM D5473. West Conshohocken, PA: ASTM.

ASTM. n.d. *Test method (analytical procedure) for determining transmissivity and storage coefficient of bounded, nonleaky, confined aquifers*. ASTM D5270. West Conshohocken, PA: ASTM.

ASTM. n.d. *Test method (field procedure) for instantaneous change in head (slug tests) for determining hydraulic properties of aquifers*. ASTM D4044. West Conshohocken, PA: ASTM.

ASTM. n.d. *Test method (field procedure) for withdrawal and injection well tests for determining hydraulic properties of aquifers*. ASTM D4104. West Conshohocken, PA: ASTM.

ASTM. n.d. *Test method for determining specific capacity and estimating transmissivity at the control well*. ASTM D5472. West Conshohocken, PA: ASTM.

ASTM. n.d. *Test method for determining subsurface liquid levels in a borehole or monitoring well (observation well)*. ASTM D4750. West Conshohocken, PA: ASTM.

Black and Veach. 1993. *Willows water district well A-6A aquifer recharge evaluation final report*. Aurora, CO: Black and Veach.

Bouwer, E. L., P. L. McCarty, H. Bouwer, and R. C. Rice. 1984. "Organic contaminant behavior during rapid infiltration of secondary wastewater at the Phoenix 23rd Avenue project." *Water Res.* 18 (4): 463–472.

Bouwer, H. 1978. *Groundwater hydrology*. New York: McGraw-Hill.

Bouwer, H. 1991. "Groundwater recharge with sewage effluent." *Water Sci. Technol.* 23 (10–12): 2099–2108.

Bouwer, H. 1997. "Role of ground water recharge and water reuse in integrated water management." *Arab. J. Sci. Eng.* 22 (1): 123–131.

Bouwer, H. 2002. "Artificial recharge of groundwater: Hydrogeology and engineering." *Hydrogeol. J.* 10 (1): 121–142.

Bouwer, H., and R. C. Rice. 1984. "Hydraulic properties of stony vadose zones." *Ground Water* 22 (6): 696–705.

Bouwer, H., and R. C. Rice. 1989. "Effect of water depth in groundwater recharge basins on infiltration rate." *J. Irrig. Drain. Eng.* 115 (4): 556–567.

Budhu, M., R. Ossai, and I. Adiyaman. 2014. "Ground movements from aquifer and recovery." *J. Hydrol. Eng.* 19 (4): 790–799.

Burns, A. W. 1985. *Hydrologic description of the Tamarack wildlife area and vicinity, Logan County, Colorado, and simulated effects of possible water management activities*. US Geological Survey Water Resources Investigations Rep. No. 85-4014. Reston, VA: USGS.

Canter, L. W. 1996. *Environmental impact assessment*. New York: McGraw-Hill.

CDWR (California Well Standards). 1998. "Water well standards (bulletins 74-81 and 74-90 combined), section III, destruction of wells." http://wwwdpla.water.ca.gov/sd/groundwater/california_well_standards/well_standards_content.html. May 13, 2016.

CSI (Construction Specifications Institute). 2014. *MasterFormat: Master list of numbers and titles for the construction industry*. Alexandria, VA: CSI.

CSIRO (Commonwealth Scientific and Industrial Research Organization). 2016. "Managed aquifer recharge." Accessed May 2016. https://research.csiro.au/mar/.

CWCB (Colorado Water Conservation Board). 2007. *SB06-193 underground water storage study*. Denver: CWCB.

Dieter, C. A., M. A. Maupin, R. R. Caldwell, M. A. Harris, T. I. Ivahnenko, J. K. Lovelace, et al. 2018. *Estimated use of water in the United States in 2015: US Geological Survey circular 1441*. Reston, VA: USGS.

Delin, G. A., and D. W. Risser. 2007. *Ground-water recharge in humid areas of the United States—A summary of ground-water resources program studies, 2003–2006*. Fact Sheet 2007-3007. Reston, VA: USGS.

Detay, M. 1996. "Rational ground water reservoir management." In *Artificial recharge of ground water II*. New York: ASCE.

Doherty, J. 2005. *PEST model-independent parameter estimation user manual,* 5th ed. Brisbane, Australia: Watermark Numerical Computing.

Drewes, J. E., C. Hoppe, D. Heil, and E. Dickenson. 2010. *Performance assessment of surface spreading operations receiving different blends of tertiary/RO treated waters*. Final Report to the Water Replenishment District of Southern California. Golden, CO: Colorado School of Mines.

Driscoll, E. G. 1986. *Groundwater and wells,* 2nd ed. St. Paul, MN: Johnson Division.

Eaton, L., J. Melady, and T. Tolan. 2009. "Successful implementation of ASR in basalt-hosted aquifers in the Pacific Northwest of the United States." *Bol. Geol. Min.* 120 (2): 131–156.

Ewing, A., J. M. Stomp, III, and K. Yuhas. 2015. "ABCWUA aquifer storage and recovery." In *Proc., National Groundwater Association Groundwater Summit*, San Antonio.

Ewing, A., K. Yuhas, and J. M. Stomp, III. 2016. "ABCWUA aquifer storage and recovery." In *Proc., National Groundwater Association Conf. on Hydrology and Water Quality in the Southwest*. Albuquerque, New Mexico.

Ewing, A., K. Yuhas, J. M. Stomp, III, and M. Bitner. 2014. "Albuquerque Bernalillo county water utility authority aquifer storage and recovery." In *Proc., National Groundwater Association Conf. on Hydrology and Water Scarcity in the Rio Grande Basin*, Albuquerque, NM.

Fetter, C. W. 2001. *Applied hydrogeology,* 4th ed. Upper Saddle River, NJ: Prentice Hall.

Fort, D. D. 2008. "ASR from a legal perspective." *Southwest Hydrol.* 7 (3): 22–23.

Fowler, L. C., ed. 1996. *Operation and maintenance of ground water facilities: Manuals and reports on engineering practice no. 86*. Reston, VA: ASCE.

Freeze, R. A., and J. A. Cherry. 1979. *Groundwater*. Englewood Cliffs, NJ: Prentice-Hall.

Gin, G. M., and N. Nutter. 2011. Aquifer storage and recovery technologies: Now and beyond. In Proc., Arizona Water Association Distribution Systems Workshop, Scottsdale and Tucson, Arizona, March 30, 2011, and April 14, 2011.

Gin, G. M., and N. Nutter. 2012. *City of Phoenix, aquifer storage and recovery well #299: 'How did you unclog that well?'*. Glendale, AZ: Radial Injection Surge Development, AZ Water Conference.

Gin, G. M., N. Nutter, and T. Morris. 2013. *City of Phoenix, aquifer storage and recovery well #299: 'How did you unclog that well?'*. San Antonio: Radial Injection Surge Development, National Groundwater Association.

Gin, G. M., A. Terrey, and C. Brand. 2011. "Indirect potable reuse: City of Phoenix's cave creek water reclamation aquifer storage and recovery well program–From conceptual plan and beyond." In *Proc., 26th Ann. Wateruse Conf.*, Phoenix, Arizona.

Gin, G. M., and D. Wrzosek. 2013. *Phoenix's ASR well: Glass beads versus silica sand filter pack: Is there a difference?* San Antonio: National Groundwater Association.

Glover, R. E. 1960. *Mathematical derivations as pertain to groundwater recharge*. Ft. Collins, CO: Agricultural Research Service, USDA.

Glover, R. E. 1974. *Transient groundwater hydraulics*. Ft. Collins, CO: Water Resources Publications.

GSI Water Solutions. 1999–2016. *City of Beaverton ASR Annual Reports, submitted to the Oregon Water Resources Department per the City of Beaverton's ASR Limited License #002*. Bend, OR: GSI Water Solutions.

Hahn, C. 2009. *Review of existing methods for addressing compliance with endangerment law under the safe drinking water act in permitting of aquifer storage and recovery wells*. Golden, CO: Environmental Science and Engineering Department, Colorado School of Mines.

Hahn, W., H. Thompson, J. Forbes, and M. Ankeny. 2003. *Comparison of alternative methods for recharge of a deep aquifer*. Denver: AWWA Research Foundation.

Haitjema, H. M. 1995. *Analytic element modeling of ground water flow*. San Diego: Academic Press.

Hallbeck, E., and K. Pederson. 1986. "The biology of Gallionella." In *Proc., Int. Symp. on Biofouled Aquifers: Prevention and Restoration*, D. R. Cullimore, ed. American Water Resources Association, Bethesda, Maryland.

Halstead, M. 2012. "Colorado Division of Water Resources; Powerpoint, Tamarack Recharge Project: How field data collection and modeling were used to determine stream impacts for the water rights case." American Groundwater Trust. http://southplatte.colostate.edu/files/AGWT/Halstead.pdf. May 15, 2013.

Hantush, M. D. 1967. "Growth and decay of groundwater mounds in response to uniform percolation." *Water Resour. Res.* 3 (1): 227–234.

Harbaugh, A. W., E. R. Banta, M. C. Hill, and M. G. McDonald. 2000. *MODFLOW-2000, the US Geological Survey modular ground-water model—User guide to modularization concepts and the Ground-Water Flow Process*. USGS Open-File Rep. No. 00–92. Reston, VA: USGS.

Heath, R. C. 1984. *Ground water regions of the United States: US Geological Survey water supply paper 2242*. Reston, VA: USGS.

Heath, R. C. 1989. *Basic ground-water hydrology: US Geological Survey water-supply paper 2220*. Denver: USGS.

Homer, C. H., J. A. Fry, and C. A. Barnes. 2012. *The national land cover database*. USGS Fact Sheet 2012–3020. Denver: USGS.

Hutchinson, A., M. Milczarek, and M. Banerjee. 2013a. "Clogging phenomena related to surface water recharge facilities." In *Clogging issues associated with managed aquifer recharge methods: IAH commission on managing aquifer recharge, Australia,* R. Martin, ed., 95–106. International Association of Hydrogeologists. Australia.

Hutchinson, A., D. Phipps, G. Rodriquez, G. Woodside, and M. Milczarek. 2013b. "Surface spreading recharge facility clogging—The Orange County water district experience." In *Clogging issues associated with managed aquifer recharge methods: IAH commission on managing aquifer recharge, Australia,* R. Martin, ed., 107–118. International Association of Hydrogeologists. Australia.

Jain, R. K., L. V. Urban, G. S. Stacey, and H. E. Balbach. 1993. *Environmental assessment.* New York: McGraw-Hill.

JCHA (John C. Halapaska and Associates). 1994. *Third quarterly report, Denver basin aquifer recharge demonstration project.* Denver: US Bureau of Reclamation.

JCHA. 1997. *Final report, Denver basin aquifer recharge demonstration project.* Denver: US Bureau of Reclamation.

Jenkins, C. T. 1970. *Computation of rate and volume of stream depletion by wells.* Reston, VA: USGS.

Johnson, A. I. 1981. "Some factors contributing to decreased well efficiency during fluid injection." In *Water for subsurface injection,* 89–101. West Conshohocken, PA: ASTM.

Kendall, C., and D. P. Krabbenhoft. 1995. "Applications of isotopes to tracing sources of solutes and water in shallow systems." In *Proc., Symp. on Groundwater Management,* R. J. Charbenau, ed., San Antonio, 390–395.

Khan, M. J., A. U. Bhatti, S. Hussain, and A. Wasiullah. 2007. "Heavy metal contamination of soil and vegetables with industrial effluents from sugar mill and tanneries." *Soil Environ.* 26 (2): 139–45.

Kimbler, O. K., R. G. Kazmann, and W. R. Whitehead. 1975. *Cyclic storage of fresh water in saline aquifers: Louisiana Water Resources Research Institute bulletin 10.* Baton Rouge, LA: Louisiana State Univ.

Knorr, D., and T. Cliett. 1985. "Proposed groundwater recharge at El Paso, Texas." In *Artificial recharge of groundwater,* T. Asano, ed., 425–480. Stoneham, MA: Butterworth Publishers.

Kruseman, G. P., and N. A. de Ridder. 1992. *Analysis and evaluation of pumping test data: ILRI publication 47,* 2nd ed. Wageningen, Netherlands: International Institute for Land Reclamation and Improvement.

Lee, G. E, and A. Jones-Lee. 1995a. "Monitoring reclaimed domestic wastewater in public parkland vegetation to reduce risks." *Water Eng. Manage.* 142: 37.

Lee, G. E, and A. Jones-Lee. 1995b. "Public health and environmental safety of reclaimed wastewater reuse." In *Proc., 7th Symp. on Artificial Recharge of Groundwater,* University of Arizona Water Resources Research Center, Tucson, Arizona.

Lee, G. E., and A. Jones-Lee. 1996. "Appropriate degree of domestic wastewater treatment before groundwater recharge and for shrubbery irrigation." In *Proc., AWWA, WEF 1996 Water Reuse Conf.,* San Diego, 929–939.

Lee, J., R. Pierrehumbert, A. Swann, and B. R. Lintner. 2009. "Sensitivity of stable water isotopic values to convective parameterization schemes." *Geophys. Res. Lett.* 36 (23): L23801.

Marley, R. 2010. "Reclaimed water aquifer storage and recovery (ASR), Rio Rancho, New Mexico." In *Proc., WESTCAS 2010 Winter Conf.,* Albuquerque, NM.

Marley, R. 2011. "MBR-AOP-BAC treatments for aquifer recharge, City of Rio Rancho, New Mexico." In *Proc., Rocky Mountain Section of the American Water Works Association Annual Conf.,* Loveland, CO.

Marley, R. 2012. *City of Rio Rancho aquifer recharge program.* Albuquerque, NM: The New Mexico Water Conservation Alliance.

Marley, B., G. Hall, V. Dholakia, T. McEvoy, and S. Porter. 2010. "Pilot scale evaluation of MBR-AOP-BAC treatments for wastewater reuse and aquifer recharge applications." In *Proc., WEFTEC 2010,* New Orleans.

McDermott, J. A., D. Avisar, T. Johnson, and J. F. Clark. 2008. "Travel times near spreading ponds: Inferences from geochemical and physical approaches." *J. Hydrol. Eng.* 13 (11): 1021–1028.

McGrane. 2013. *Feasibility report—Expansion of an existing (confidential) underground storage project in the Phoenix basin, Arizona.*

Miller, D. W. 1980. *Waste disposal effects on groundwater.* Berkeley, CA: Premier Press.

Miller, M. M., and M. Shirzaei. 2015. "Spatiotemporal characterization of land subsidence and uplift in Phoenix using InSAR time series and wavelet transforms." *J. Geophys. Res. Solid Earth* 120 (8): 5822–5842.

Moore, S. J., J. T. Kay, and J. M. Stomp, III. 2007. "New Mexico's first artificial recharge project." In *Proc., 2007 AWRA National Conf.,* Albuquerque, NM.

Moore, S. J., and J. M. Stomp, III. 2009. "The Bear Canyon recharge demonstration project." In *Water, natural resources, and the urban landscape: The Albuquerque Region: New Mexico Bureau of Geology and Mineral Resources 2009 decision-makers field guide,* L. G. Price, D. Bland, P. S. Johnson, and S. D. Connell, eds., 34. Socorro, NM: New Mexico Bureau of Geology and Mineral Resources.

Murtaza, G., A. Ghafoor, and M. Qadir. 2008. "Accumulation and implications of cadmium, cobalt and manganese in soils and vegetables irrigated with city effluent." *J. Sci. Food Agric.* 88 (1): 100–107.

Nellor, M. H., R. B. Baird, and L. R. Smith. 1984. *Summary of health effects study: Final report.* Whittier, CA: County Sanitation Districts of Los Angeles County.

NGWA (National Groundwater Association). 2014. "Best suggested practices for aquifer storage and recovery." Accessed December 17, 2018. https://www.ngwa.org/publications-and-news/industry-resource-library/Industry-best-suggested-practices.

Nikos, S., P. Ioannis, L. Constantinos, T. Paraskevas, K. Anastasia, and H. K. Charalambos. 2016. "Land subsidence rebound detected via multi-temporal InSAR and ground truth data in Kalochori and Sindos regions, Northern Greece." *Eng. Geol.* 209 (15): 175–186.

NRC (National Research Council). 1994. *Groundwater recharge using waters of impaired quality.* Washington, DC: National Academies Press.

NRC. 2008. *Prospects for managed underground storage of recoverable water.* Washington, DC: National Academies Press.

NRC. 2013. *Induced seismicity potential in energy technologies.* Washington, DC: National Academies Press.

NRMMC, EPHC, and NHMRC (Natural Resource Management Ministerial Council, Environment Protection and Heritage Council, and National Health and Medical Research Council). 2009. "Australian guidelines for water recycling, managing health and environmental risks, Volume 2C—Managed aquifer recharge." http://www.environment.gov.au/resource/national-water-quality-management-strategy-australian-guidelines-water-recycling-managing-1. Accessed May 24, 2010.

NWQMS (National Water Quality Management Strategy). 2009. "Australian guidelines for water quality monitoring and reporting." Australian Department of the Environment. https://www.environment.gov.au/water/quality/publications/nwqms-australian-guidelines-water-quality-monitoring-reporting. Accessed May 4, 2010.

Orange County Water District. 2018. *Groundwater replenishment system 2017 annual report.* Prepared for the California Regional Water Quality Control Board, Santa Ana Region, Order No. R8-2004-0002, as amended by Order Nos. R8-2008-0058, R8-2014-0054, and R8-2016-0051. Fountain Valley, CA: Orange County Water District.

OWRB (Oklahoma Water Resources Board). 2010. *Final report aquifer recharge pilot project site evaluation.* Senate Bill 1410

Aquifer Recharge Work Group, Oklahoma Comprehensive Water Plan. Oklahoma City: OWRB.
Pavelic, P., P. J. Dillon, and C. T. Simmons. 2002. "Lumped parameter estimation of initial recovery efficiency during aquifer storage and recovery." In *Management of Aquifer Recharge for Sustainability: Proc., 4th Int. Symp. on Artificial Recharge (ISAR4)*, P. J. Dillon, ed., Adelaide, Australia.
Pavelko, M. T., J. Hoffmann, and N. A. Damar. 2006. *Interferograms showing land subsidence and uplift in Las Vegas Valley, Nevada, 1992–99.* USGS Scientific Investigations Rep. No. 2006-5218. Reston, VA: USGS.
Pettyjohn, W. A., J. Studlick, R. C. Bain, and J. H. Lehr. 1979. *A ground-water quality atlas of the United States.* Washington, DC: National Demonstration Water Project.
Piper, A. M. 1944. "A graphic procedure in the geochemical interpretation of water-analyses." *Am. Geophys. Union* 25 (6): 914–928.
Poeter, E. P., M. C. Hill, E. R. Banta, S. Mehl, and S. Christiansen. 2005. *UCODE_2005 and six other computer codes for universal sensitivity analysis, calibration, and uncertainty evaluation: USGS techniques and methods 6-A11.* Reston, VA: USGS.
Pyne, R. D. G. 1995. *Groundwater recharge and wells: A guide to aquifer storage recovery.* Boca Raton, FL: Lewis Publishers.
Pyne, R. D. G. 2005. *Aquifer storage recovery: A guide to groundwater recharge through wells.* Gainesville, FL: ASR Press.
Reese, R. S. 2002. *Inventory and review of aquifer storage and recovery in southern Florida.* USGS Water Resources Investigation Rep. No. 02-4036. Reston, VA: USGS.
Reilly, T. E., and A. W. Harbaugh. 2004. *Guidelines for evaluating ground-water flow models.* USGS Scientific Investigations Rep. No. 2004-5038. Reston, VA: USGS.
Reilly, T. E., K. F. Dennehy, W. M. Alley, and W. L. Cunningham. 2008. *Ground-water availability in the United States:* USGS Circular 1323. Reston, VA: USGS.
Roscoe Moss Company. 1982. *A guideline to water well casing and screen selection.* Los Angeles: Rosco Moss.
Roscoe Moss Company. 1985. *The engineers' manual for water well design.* Los Angeles: Rosco Moss.
Roscoe Moss Company. 1990. *Handbook of ground water development.* New York: Wiley.
Sheffield, C. W., C. Rivero-deAguilar, and K. McCann. 1995. "Drainage wells—A dual purpose." *Florida Water Resources J.* March.
Sheng, Z. 2005. "An aquifer storage and recovery system with reclaimed wastewater to preserve native groundwater resources in El Paso, Texas." *J. Environ. Manage.* 75 (4): 367–377.
Sheng, Z., J. Arthur, J. Restrepo, and R. Allen-King. 2007. "On recovery efficiency of managed underground storage of recoverable water systems." In *Proc., 6th Biennial Int. Symp. on Managed Aquifer Recharge*, Phoenix, 84–95.
Shiklomanov, I. 1993. "World fresh water resources." In *Water in crisis: A guide to the world's fresh water resources,* P. H. Gleick, ed. New York: Oxford University Press.
Shrier, C. J., D. Dunford, and J. Altenhofen. 1999. "The development of a multi-criteria managed recharge facility to augment instream flows and restore protected species habitat." In *Proc., Watershed Management to Protect Declining Species.* AWRA's 1999 Ann. Water Resources Conf., Seattle.
Sloss, E. M., S. A. Geschwind, D. F. McCaffrey, and B. R. Ritz. 1996. *Groundwater recharge with reclaimed water: An epidemiological assessment in Los Angeles County,* 1987–1991. Santa Monica, CA: Rand.
Sterrett, R. J. 2007. *Groundwater and wells,* 3rd ed. St. Paul, MN: Johnson Division.
Swiezckowski, D., S. Morris, and T. Carpenter. 2007. "An overview of Arizona's recharge program." In *Proc., 6th Int. Symp. on Managed Aquifer Recharge, ISMAR6*, P. Fox, ed., 366–378. Phoenix, Arizona.
SWRCB (State Water Resources Control Board, State of California). 2015. *Regulations related to recycled water.* Sacramento, CA: SWRCB.
Taylor, G. F. 1993. *Inventory of drainage wells and potential sources of contaminants to drainage-well inflow in Southwest Orlando, Orange County, Florida.* USGS Water-Resources Investigations Rep. No. 93-4061. Denver: USGS.
Theis, C. V. 1941. "The effect of a well on the flow of a nearby stream." *Eos, Trans. Am. Geophys. Union* 22 (3): 734–738.
Theis, C. V. 1952. *The relation between the lowering of the piezometric surface and the rate and duration of discharge of a well using ground water storage: USGS ground water notes (hydraulics) no. 5.* Reston, VA: USGS.
Todd, D. K., and L. W. Mays. 2005. *Ground water hydrology,* 3rd ed. New York: Wiley.
Topper, R., P. E. Barkmann, D. A. Bird, and M. A. Sares. 2004. *Artificial recharge of ground water in Colorado—a statewide assessment: Environmental geology 13.* Denver: Colorado Geological Survey, Dept. of Natural Resources.
Thurman, E. M. 1979. "Isolation, characterization, and geochemical significance of humic substances from groundwater." Ph.D. thesis, Dept. of Geological Sciences, Univ. of Colorado, Boulder.
UN (United Nations). 1975. *Groundwater storage and artificial recharge: Department of social and economic affairs, water series no. 2.* New York: UN.
USBLM (US Bureau of Land Management). 2006. "Areas of critical environmental concern." Accessed November 1, 2006. http://www.co.blm.gov/metadata/co_acec.htm.
USEPA. 1992. *Guidelines for water reuse: Manual EPA 625/R-92/004.* Washington, DC: USEPA.
USEPA. 1999. *The class V underground injection control study: Chapter 21, aquifer recharge and aquifer storage and recovery wells: Manual EPA/816-R-99-014u.* Washington, DC: USEPA.
USEPA. 2013. "Letter from P. Grevatt, Director of EPA office of ground water and drinking water to M. Thomasson, Director of Division of water resource management, Florida Department of Environmental Protection." September 27.
USEPA. 2016. "Aquifer recharge and aquifer storage and recovery." Accessed June 17, 2016. https://www.epa.gov/uic/aquifer-recharge-and-aquifer-storage-and-recovery.
USGS. 2006. "National hydrography dataset." Accessed March 1, 2007. https://www.usgs.gov/core-science-systems/ngp/national-hydrography.
Vail, R. 2012. "Colorado parks and wildlife: Powerpoint, Tamarack Ranch State Wildlife Area Managed Groundwater Recharge Project."American Groundwater Trust Website. Accessed February 21, 2017. http://southplatte.colostate.edu/files/AGWT/Vail.pdf.
Wallace, C. 2012. "Colorado attorney general: Powerpoint, negotiated resolution of the Tamarack recharge project." American Groundwater Trust. Accessed February 17, 2017. http://southplatte.colostate.edu/files/AGWT/Wallace.pdf.
Winter, T. C., J. W. Harvey, O. L. Franke, and W. M. Alley. 1998. *Ground water and surface water: A single resource: USGS circular 1139.* Denver: USGS.
Wolf, C., B. Marley, and A. Ewing. 2016. "Applying geochemistry to managed aquifer recharge projects." In *Proc., National Groundwater Association Conf. on Hydrology and Water Quality in the Southwest,* Albuquerque, NM.

Yadav, R. K., B. Goyal, R. K. Sharma, S. K. Dubey, and P. S. Minhas. 2002. "Post-irrigation impact of domestic sewage effluent on composition of soils, crops and ground water—A case study." *Environ. Int.* 28 (6): 481–486.

Zhang, M., and T. J. Burbey. 2016. "Inverse modelling using PS-InSAR data for improved land subsidence simulation in Las Vegas Valley, Nevada." *Hydrol. Process.* 30 (24): 4494–4516.

Zheng, C. 1990. *MT3D: A modular three-dimensional transport model for simulation of advection, dispersion and chemical reactions of contaminants in groundwater systems.* Tuscaloosa, AL: Univ. of Alabama.

INDEX

Figures are indicated by *f*; tables are indicated by *t*.